Ecological Studies

Analysis and Synthesis

Volume 255

Ecological Studies is Springer's premier book series treating all aspects of ecology. These volumes, either authored or edited collections, appear several times each year. They are intended to analyze and synthesize our understanding of natural and managed ecosystems and their constituent organisms and resources at different scales from the biosphere to communities, populations, individual organisms and molecular interactions. Many volumes constitute case studies illustrating and synthesizing ecological principles for an intended audience of scientists, students, environmental managers and policy experts. Recent volumes address biodiversity, global change, landscape ecology, air pollution, ecosystem analysis, microbial ecology, ecophysiology and molecular ecology.

Xoaquín Moreira • Luis Abdala-Roberts
Carla Vázquez-González
Editors

Effects of Plant Diversity on Plant-Herbivore Interactions

Springer

Editors
Xoaquín Moreira
Misión Biológica de Galicia (CSIC)
Pontevedra, Spain

Carla Vázquez-González
Misión Biológica de Galicia (CSIC)
Pontevedra, Spain

Luis Abdala-Roberts
Universidad Nacional Autónoma de México, Escuela Nacional de Estudios Superiores-Unidad Mérida
Ucú, Yucatán, Mexico

ISSN 0070-8356 ISSN 2196-971X (electronic)
Ecological Studies
ISBN 978-3-032-26460-2 ISBN 978-3-032-26461-9 (eBook)
https://doi.org/10.1007/978-3-032-26461-9

This Springer imprint is published by the registered company Springer Nature Switzerland AG
The registered company address is: Gewerbestrasse 11, 6330 Cham, Switzerland

To my daughters, Olivia and Marina, whose laughter brightens every day; to my wife, Silvia, for her endless support; and to my parents, Joaquín and Aurora, for their constant encouragement, with deepest gratitude.

—Xoaquín Moreira

To ECO-EVO research group and all its members, including the technical staff, without whom our research would so often not be possible.

—Carla Vázquez-González

To my dear friend Jorge Berny, for the original inspiration to plant a tree diversity experiment, and to my family for their support and love for nature which undoubtedly shaped my views on life and scientific career.

—Luis Abdala-Roberts

Foreword

The study of plant biodiversity sits at the nexus between community and ecosystem ecology. Long before biodiversity was a conservation imperative, ecologists recognized that plant communities structure ecosystems. Plants regulate primary productivity, influence nutrient cycling, and determine the architecture of food webs. What emerged more gradually was the recognition that plant diversity governs trophic interactions—and through them, ecosystem function.

Early plant–herbivore theory focused on pairwise consumer–resource dynamics, with density dependence being foundational: herbivore populations increase with host abundance, and plant damage intensifies as host plants become locally common. These principles explained outbreak dynamics and population regulation. They also implied a broader insight. If herbivory depends on host density, then plant neighbourhood composition must shape herbivore pressure. Herbivory is not solely a function of plant traits; it is mediated by ecological context.

These insights were formalized in the concepts of associational resistance and associational susceptibility. Plants growing among heterospecific neighbours often experience reduced herbivory relative to monocultures, though in some cases neighbours increase damage. Mechanisms such as host dilution, disrupted foraging efficiency, and altered patch residence times demonstrated that plant community composition reshapes consumer behaviour. Density dependence became embedded within diversity. Apparent competition then extended this framework. When two plant species share a common herbivore, one species can indirectly elevate herbivore pressure on the other by sustaining higher enemy densities. Communities thus become linked through shared consumers. Herbivory reflects not only local abundance but also network structure.

Plant apparency theory introduced an evolutionary dimension, where trait evolution cannot be separated from neighbourhood structure. Long-lived, abundant, or spatially predictable plants—those most apparent to herbivores—were predicted to invest differently in defence than ephemeral or rare species. In this way, optimal defence allocation must reflect exposure.

Advances in chemical ecology reinforced this contextual view. Defensive traits, particularly secondary metabolites, are often over dispersed within plant communities. Neighbouring species frequently differ markedly in chemical profiles, creating heterogeneity that can disrupt herbivore host location and feeding. Diversity generates a mosaic of plant defensive landscapes through which herbivores must navigate. Density-dependent feedbacks unfold within chemically structured communities.

Incorporating a third trophic level further expands this framework. The Enemies Hypothesis proposed that increased plant diversity enhances predator and parasitoid abundance, thereby suppressing herbivores. Diverse plant assemblages provide structural complexity, alternative prey, refugia, and favourable microclimates that recruit natural enemies. In this tri-trophic perspective, plant diversity regulates herbivory indirectly through predator enhancement.

These mechanisms do not operate independently. Associational resistance may arise from host dilution, altered apparency, chemical heterogeneity, predator recruitment, or combinations thereof. Apparent competition may strengthen or weaken depending on enemy responses. Density dependence unfolds within a network of bottom-up and top-down forces.

Chemically mediated interactions and signalling among plants, and between plants and predator and microbes add a further layer of complexity to these interactions. Plants emit herbivore-induced volatiles that function as defensive signals and ecological cues. These "cry for help" volatiles attract predators and parasitoids. Belowground, root exudates and volatile signals alter microbial communities and influence neighbouring plant responses. These signalling processes occur within diverse assemblages. In species-rich communities, plant volatile plumes overlap and interact. Neighbouring plants may perceive cues and upregulate defences preemptively, synchronizing resistance across individuals. Alternatively, chemical heterogeneity may mask or dilute cues, complicating herbivore host detection. Predator recruitment via volatiles may be amplified in structurally complex mixtures that support diverse enemy communities. Volatile-mediated signalling thus provides a mechanistic bridge linking trait diversity, associational effects, and tri-trophic regulation. It extends biodiversity effects into chemical signalling networks that operate invisibly but shape trophic dynamics. Whether overlapping volatile emissions amplify or dampen predator attraction, and how these processes scale in species-rich forests, remain open questions.

Large-scale biodiversity–ecosystem function experiments have created opportunities to integrate these theories. Early work emphasized productivity and stability. Subsequent multitrophic analyses demonstrated that plant diversity frequently reduces herbivory through host dilution, trait-mediated defence shifts, predator recruitment, and context-dependent signalling. In many systems, strengthened top-down control contributes to enhanced ecosystem performance.

The chapters in this volume synthesize this progression. Trait-based approaches clarify how plant chemistry, morphology, and nutritional attributes shift in mixtures. Empirical studies across grasslands, forests, and urban systems evaluate bottom-up and top-down mechanisms. Long-term analyses reveal how diversity–herbivory relationships change over time. Predator-focused research examines when increased enemy abundance translates into effective trophic control.

Global change heightens the urgency of this synthesis. Drought alters plant physiology and volatile emissions. Warming shifts herbivore phenology and predator behaviour. Nutrient enrichment modifies defensive allocation. Whether plant diversity buffers ecosystems through chemical heterogeneity, synchronized defence induction, or sustained predator assemblages remains uncertain. Predicting resilience requires integrating these mechanisms.

The intellectual trajectory traced here—from density-dependent consumer dynamics to associational theory, apparent competition, apparency theory, trait overdispersion, tri-trophic regulation, and volatile-mediated communication—reflects a steady expansion of ecological perspective. Pairwise interaction theory has matured into a network-based understanding of biodiversity-mediated trophic regulation operating across spatial scales and through both visible and chemical pathways.

This volume represents a synthesis of that trajectory. It integrates foundational theory with contemporary mechanistic insight across ecosystems and latitudes. It demonstrates that plant diversity shapes not only species composition but also the chemical, behavioural, and trophic networks that determine ecosystem function. At a time of accelerating biodiversity loss and climate change, such integration is necessary. Ecosystems are increasingly novel assemblages. Understanding how diversity restructures interaction networks—aboveground and belowground, direct and indirect—is central to predicting their stability and resilience. Plant diversity is not merely a count of species. It is a structural force that reorganizes ecological networks—from roots to leaves, from herbivores to predators, and through the volatile signals that connect them.

Irvine, CA, USA
18 February 2026

Kailen Mooney

Preface

Luis and Xoaquín met in the group of Professor Kailen Mooney at the University of California, Irvine, in 2012. Coincidentally, both began working on the effects of plant diversity on species interactions around the same time, before knowing each other. Before moving to California in 2012, Xoaquín had just finished a study in Galicia, Spain, where he established experimental plots testing the effects of pine seedling diversity on interactions with ants and insect herbivores. He further manipulated ant presence across seedling plots to demonstrate—in what remains a groundbreaking study—that predatory ants mediate pine diversity effects on seedling growth and establishment. The implications for pine forest conservation and restoration in this region were noteworthy. Around the same time, in 2011, Luis obtained funding to establish the first tree diversity experiment in Mexico (the third in Latin America at the time) in southern Yucatán, Mexico (the UADY–UNAM Tree Diversity Experiment), examining effects on insect herbivory and tri-trophic interactions. The system is now 15 years old and has provided key insights into tropical tree diversity effects on plant–insect interactions, resulting in more than 20 publications over this period, including studies on several tropical tree insect pests of major economic importance. Interestingly, the year after they met, Xoaquín obtained funding to work in this system, investigating tree diversity effects on tree defences as a novel mechanism linking producer diversity to trophic interactions. Since then, they have continued collaborating in this system, obtaining additional funding that has resulted in several publications, as well as a well-cited synthesis paper on plant diversity mechanisms affecting biotic interactions. More recently, in yet another academic connection, Carla joined Kailen Mooney's group and began collaborating extensively with Luis and Xoaquín. One of their most important research themes has focused on the effects of insectivorous birds on herbivory and plant defences, including novel work from the UADY–UNAM system examining tree diversity effects on the stability of, and predatory activity by, this bird feeding guild. She later went on to lead an international research collaboration (as part of the Tree Diversity Network) testing tree diversity effects on predatory activity by birds and predatory arthropods across systems in Europe, North America (including Yucatán), and Asia, providing the first tests of biogeographic patterns associated with tree diversity effects on predator top-down control. After more than a decade working together, sharing projects and students, and discussing different perspectives from our research, we have gained important insights into the mechanisms by which plant diversity affects trophic interactions. The motivation to put together this book arose from the many discussions the editors have had in recent years. One of their main conclusions is that, despite the recognized importance of plant diversity and herbivory for ecosystem functioning, we still lack a robust understanding of how changes in herbivory—through bottom-up or top-down processes—mediate plant diversity–function relationships. In this sense, the present volume seeks to advance our understanding of these linkages, ultimately providing mechanistic knowledge on a central question in ecology: How does plant community structure, with diversity as one of its main properties, influence ecosystem functioning via biotic interactions across trophic levels? The implications are wide-ranging, from improved mechanistic explanations of population stability, food web regulation, and ecosystem service delivery, to the further development of applied ecological sciences that seek to use plant diversity as a tool for conserving and designing more resilient ecosystems in both natural and managed environments.

Pontevedra, Spain Xoaquín Moreira
Ucú, Yucatán, Mexico Luis Abdala-Roberts
Pontevedra, Spain Carla Vázquez-González

Acknowledgements

First of all, we would like to sincerely thank all the authors who enthusiastically contributed to this book. Their dedication and hard work in writing the chapters made this volume possible. This book simply would not exist without their invaluable contributions. We are also deeply grateful to the staff of Springer Nature—Lars Koerner, Peggy Papandreou, and Bibhuti Bhusan Sharma—for their support and guidance throughout the entire editorial process. Their expertise and assistance were essential in bringing this book to completion.

Contents

Introduction: Effects of Plant Diversity on Plant-Herbivore Interactions

1

Xoaquín Moreira, Carla Vázquez-González, and Luis Abdala-Roberts

Abstract

The causal link between plant diversity and ecosystem functioning represents a cornerstone of ecology. Experimental research has also convincingly shown that plant diversity shapes trophic interactions such as herbivory and predation, ultimately contributing to food web regulation. Yet, these facets of biodiversity research have developed somewhat independently, limiting consensus on the role of trophic interactions—particularly herbivory, which directly affects plant productivity and other functions—in linking plant diversity to higher trophic levels and overall food web dynamics. This volume aims to address this gap by synthesizing current knowledge on how plant diversity influences herbivory, the underlying mechanisms, and the implications for ecosystem processes. Drawing on long-term studies in natural and experimental grasslands, forests, and urban vegetation, results indicate that plant diversity can reduce and stabilize herbivore pressure through multiple mechanisms, among which enhanced top-down effects from higher trophic levels play a prominent role. Notably, evidence has accumulated rapidly in forests, and research is beginning to explore these dynamics in urban ecosystems as well, with applied studies increasingly guiding the use of plant diversity for pest control, microclimate regulation, and resistance to disturbances across habitat types. Despite remaining challenges, a deeper understanding of plant diversity effects on trophic interactions promises to strengthen ecological theory and advance the use of plant diversity as a nature-based solution to enhance ecosystem services and biodiversity conservation.

X. Moreira (✉) · C. Vázquez-González
Misión Biológica de Galicia (MBG-CSIC), Pontevedra, Galicia, Spain

L. Abdala-Roberts
Universidad Nacional Autónoma de México, Escuela Nacional de Estudios Superiores-Unidad Mérida, Ucú, Yucatán, Mexico

X. Moreira et al. (eds.), *Effects of Plant Diversity on Plant-Herbivore Interactions*, Ecological Studies 255, https://doi.org/10.1007/978-3-032-26461-9_1

1.1 Plant Diversity, Ecosystem Function, and Species Interactions

Plant species diversity plays a central role in shaping the structure and functioning of ecological communities, acting as a stabilizing force that maintains ecosystem processes over time (Tilman et al. 1996; Hector et al. 2010). These effects are increasingly recognized as being mediated by plant–consumer interactions, including herbivory, host-pathogen interactions, and predation, as well as plant–pollinator and plant–microbe mutualisms (Duffy et al. 2007; Scherber et al. 2010; Hambäck et al. 2014; Schuldt et al. 2019; Jactel et al. 2021; Staab et al. 2021). For plant–enemy interactions, various mechanisms have been shown to reduce density-dependent herbivore or pathogen attack on focal host plants in diverse communities (reviewed by Moreira et al. 2016; Jactel et al. 2021). At the same time, positive effects on plant mutualists often arise from increased diversity of plant-based resources, such as floral resources and other plant foods, which benefit pollinators and natural enemies and, in turn, enhance plant performance (Fornoff et al. 2019; Staab et al. 2023). Collectively, research on species interactions across trophic levels has the potential to fill key knowledge gaps and contribute in novel and insightful ways to our understanding of how plant diversity shapes communities and ecosystems.

1.2 Plant Diversity Controls Over Herbivory

Plant diversity is an important driver of herbivory, a key interaction that regulates plant population dynamics, community structure, and ecosystem functions such as primary productivity and nutrient cycling (Barbosa et al. 2009; Moreira et al. 2016; Bröcher et al. 2023). Understanding how plant diversity influences herbivory is therefore crucial for elucidating controls over ecosystem function, informing ecosystem management—including restored and agricultural systems—and conserving biodiversity across trophic levels. Within this context, plant diversity has repeatedly been shown to affect herbivore abundance and behavior, thereby altering the magnitude and spatiotemporal patterns of herbivory (Moreira et al. 2016; Jactel et al. 2021). Plant diversity often reduces herbivory (Barbosa et al. 2009; Underwood et al. 2014), as the presence of multiple plant species in diverse communities can interfere with herbivore host location and feeding efficiency (Atsatt and O'Dowd 1976; Castagneyrol et al. 2014; Ruttan and Lortie 2015). Additionally, the mere reduction in density of host plant species with increasing diversity can reduce host–enemy encounter rates or patch residence times (Hambäck et al. 2014). In some cases, however, plant diversity has been shown to increase herbivory by generalist herbivores that benefit from increased resource diversity (White and Whitham 2000; Barbosa et al. 2009). Accordingly, negative effects of plant diversity on herbivory are expected to be more likely and stronger in ecosystems with a greater representation of dominant specialized herbivores.

Plant diversity also influences herbivory indirectly through enhanced top-down effects by natural enemies (Letourneau et al. 2011). Higher plant diversity can enhance the abundance, diversity, and activity of predators and parasitoids, including birds, spiders, ants, and other predatory insects (reviewed by Staab and Schuldt 2020). These effects arise from factors such as increased availability of refugia from intraguild predation, access to alternative prey, and improved microclimatic conditions that promote predator survival and activity (Moreira et al. 2016; Jactel et al. 2021). Recent work has shown that enhanced top-down pressure in plant mixtures has in turn been shown to contribute to positive plant diversity–productivity relationships (Albert et al. 2010; Li et al. 2023). Teasing apart the relative contributions of direct and indirect (predator-mediated) effects on herbivory remains a key challenge, and in the latter case, a better understanding of changes in trophic cascade strength as a function of predator and herbivore traits is also needed (Abdala-Roberts and Moreira 2025).

Finally, environmental context influences direct and indirect pathways of plant diversity effects on herbivory. Abiotic factors such as soil resource availability, climate, and disturbance regimes can modify associational resistance or susceptibility dynamics through changes in plant physical or chemical traits responsive to such factors (e.g., Castagneyrol et al. 2018; Bröcher et al. 2025) and affecting herbivores and predators. Likewise, anthropogenic disturbances such as habitat fragmentation, land-use change, or climate warming may also directly affect herbivore or predator communities (e.g., Kneitel and Chase 2004; Lee et al. 2024). An open question is whether tree diversity buffers predator communities against harsher climatic conditions, such as elevated temperatures or increased drought frequency (see Vázquez-González et al. 2024), contributing to stability in top-down control under changing conditions.

1.3 Content Overview

Drawing on long-term research from diverse ecosystems worldwide—including experimental grasslands, natural and managed forests, and urban plant communities—this volume advances our understanding of the causal relationships between plant diversity, trophic interactions (herbivory and predation), and ecosystem processes. The chapters explore how these interactions link and explain diversity–function outcomes, highlighting the mechanisms underlying these relationships. Particular attention is given to plant traits that both predict and respond to diversity-driven bottom-up effects, as well as community-level processes shaping herbivory and ecosystem functioning. A recurring theme is the role of extrinsic factors, such as abiotic conditions and management practices, in modulating the effects of plant diversity on herbivory. Collectively, this research advances biodiversity–function studies.

Chapter 2, by Felix and Koricheva, offers a trait-based perspective on how plant diversity—both intra- and interspecific—influences herbivory and interactions with natural enemies through changes in chemical, nutritional, and morphological plant traits. They distinguish between trait-mediated bottom-up effects arising directly from increased plant diversity—such as changes in trait composition or complexity—and those emerging indirectly via plant–plant interactions in mixtures, including competition, facilitation, or niche partitioning, which alter trait expression relative to monocultures. Such interactions can create trade-offs in resource allocation, affecting plant growth versus defense traits. The authors also examine trait changes that indirectly influence herbivory by enhancing top-down control by predators. They find that effects of interspecific diversity on traits regulating herbivore abundance tend to be more consistent than those of intraspecific diversity and highlight mechanisms underlying both. These insights provide a foundation for advancing theory and empirical research on associational resistance and the direct and indirect pathways by which plant traits shape plant diversity effects on herbivores. Lastly, they emphasize the importance of identifying key traits—such as volatile compounds for insect herbivores or spines for vertebrate herbivores—and continuing comparisons of intra- versus interspecific diversity effects.

Chapter 3, by Bröcher et al., reviews the effects of plant diversity on insect herbivory in experimental temperate grasslands, focusing on how plant species richness influences invertebrate herbivore abundance and feeding behavior (Fig. 1.1). The authors highlight multiple mechanisms, including resource dilution, associational resistance, and trait-mediated interactions. Despite temperate grasslands

Fig. 1.1 A view of the Jena Experiment, a long-term grassland biodiversity study established in Jena, Germany, in 2002. (Photo credit: Anne Ebeling)

being among the most extensively studied systems in plant diversity research, the authors note that these ecosystems remain largely unexamined outside Europe and North America, leaving open questions about whether observed mechanisms vary in strength or relative importance in other grassland types, such as tropical systems, even when vegetation structure or plant–consumer functional groups are similar. In contrast to more recent experimental tree diversity research, a considerable gap remains in understanding multi-trophic interactions driving plant diversity effects on herbivory in grasslands. This chapter also highlights management practices as a key contingency factor influencing the magnitude and direction of plant diversity effects on herbivore communities and herbivory rates, offering valuable insight into the complex interactions among plant diversity, disturbances, and ecosystem functioning.

Building on these findings, Chap. 4, by Wang et al., synthesizes research across different plant community types, including forests, focusing on how plant functional and phylogenetic diversity influence herbivory from a temporal perspective. Common themes across systems include plant phenological asynchrony, temporal shifts in herbivore communities, and the moderating effects of extrinsic factors, all shaping the relationship between plant diversity and ecosystem functioning over time. The authors emphasize the importance of long-term studies, especially in systems with long-lived plant species, to test ecological theory on temporal mechanisms of diversity–stability relationships and to generate knowledge for more effective management practices. Understanding these dynamics allows more accurate predictions of how plant diversity affects herbivory, ecosystem stability, and resilience through trophic interactions.

Chapter 5, by Kempel and Allan, synthesizes how global change influences herbivory and plant–herbivore interactions. They examine responses of herbivory rates and patterns to disturbances such as warming, altered precipitation regimes, nitrogen enrichment, and land-use intensification. The authors explore how plant diversity interacts with these drivers to shape herbivory and broader ecosystem functions, considering direct effects on herbivores and linking these effects to plant community structure. They provide insights into how maintaining or enhancing plant diversity can buffer ecosystems against the impacts of global change, emphasizing that plant diversity is a key determinant of herbivory under current conditions and a potential tool for mitigating environmental disturbances in both natural and managed systems. Chapter 6, by Castagneyrol et al., presents a case study on the interaction between global change drivers and plant diversity effects on herbivory, synthesizing a decade of research from a large-scale temperate tree diversity experiment in southern France. This chapter provides mechanistic insights into how drought influences plant physiology, resource allocation, and defensive traits, and how tree diversity interacts with drought by affecting water-use efficiency through plant–plant interactions (e.g., resource partitioning) and microclimate modification (e.g., via canopy filling). The authors also consider multitrophic interactions as a mechanism by which mixed-species tree stands can buffer plant communities against herbivory under drought conditions.

Finally, Chaps. 7, 8, and 9 focus on plant diversity effects on predators—ants and insectivorous vertebrates—and their links to herbivory and ecosystem functioning in natural and experimental plant communities. Chapter 7, by Castagneyrol et al., addresses plant diversity effects on herbivory in urban forests, providing key insights into underlying mechanisms, particularly multitrophic interactions. Notably, indirect pathways mediated by top-down control appear relatively consistent in urban forests, whereas mechanisms such as host plant dilution or associational resistance are less predictable. The authors suggest that these patterns reflect intrinsic qualities of urban ecosystems, which may enhance direct pathways to herbivory relative to predator-mediated effects, and call for further research to clarify the conditions under which different mechanisms predominate.

Chapter 8, by Moreira et al., examines plant diversity effects on ant communities and ant–plant interactions. Although studies are limited, results are promising given the key ecological roles ants play as predators and scavengers in multitrophic networks and their responsiveness to changes in plant diversity. While plant diversity effects on ant-mediated trophic control of herbivory are likely important, nuances arise due to variation in ant functional traits and natural history. Traits such as defensive behavior, rapid numerical responses, and group foraging may enhance herbivore suppression in diverse plant communities, whereas ant–herbivore mutualisms (Fig. 1.2) or interference with other predators may weaken top-down control. Understanding the contexts in which positive plant diversity effects on ants translate into effective trophic control can inform strategies for insect pest management using this underexploited predator group.

Chapter 9 focuses on plant diversity effects on insectivorous vertebrates, primarily birds (Fig. 1.3), and, to a lesser extent, bats. Evidence shows consistent positive effects of tree diversity on the abundance and diversity of these groups, often

Fig. 1.2 Ants tending *Cinara pinimaritimae* aphids on *Pinus pinaster* in a plant species diversity experiment in Spain. (Photo credit: Xoaquín Moreira)

Fig. 1.3 Top panel: a view of the UADY Tree Diversity Experiment, a long-term tree biodiversity study established in 2011 in the Yucatán Peninsula, Mexico. (Photo credit: Luis Abdala-Roberts). Bottom panel: two individuals of *Passerina ciris*, an insectivorous passerine bird native to Central America, Mexico, and the southern United States, were observed in this experimental plantation. (Photo credit: Yaneli May-Uc)

correlating with enhanced herbivore suppression. However, as with ants, the scarcity of studies measuring both herbivory and plant performance limits strong conclusions about causality. The authors highlight the need for more long-term, manipulative studies, including predator exclusions, to disentangle mechanisms and quantify the contribution of vertebrate predators to top-down control of herbivores. Collectively, these predator-focused chapters underscore the importance of integrating predator diversity and behavior into our understanding of plant diversity–herbivory–ecosystem function relationships.

Overall, the book's contribution and originality stem from its consideration of diverse community types across latitudes, spanning grasslands to forests and temperate to tropical ecosystems, as well as its multifaceted approach integrating trait-based functional ecology, plant defense theory, multitrophic interactions, food web ecology, and explicit consideration of global change drivers. The volume also explores applied perspectives, including work in plantations and urban systems, aiming to identify and disentangle common mechanisms across habitat types. Crucially, by emphasizing the role of species interactions, it clarifies the mechanisms linking plant diversity to ecosystem function while providing actionable knowledge to inform management strategies that enhance resilience in both natural and managed systems under different contexts of global change.

References

Abdala-Roberts L, Moreira X (2025) Effects of plant diversity on predator-predator interactions in terrestrial ecosystems. In: Gontijo L, Wajnberg E (eds) The ecology of plant diversity: fundamentals and applications. Springer

Albert HC, Thuiller W, Gilles N, Douzet R, Aubert S, Lavorel S (2010) A multi-trait approach reveals the structure and the relative importance of intra- vs. interspecific variability in plant traits. Funct Ecol 24:1192–1201

Atsatt PR, O'Dowd DJ (1976) Plant defense guilds. Science 193:24–29

Barbosa P, Hines J, Kaplan I, Martinson H, Szczepaniec A, Szendrei Z (2009) Associational resistance and associational susceptibility: having right or wrong neighbors. Annu Rev Ecol Evol Syst 40:1–20

Bröcher M, Ebeling A, Hertzog L, Roscher C, Weisser W, Meyer ST (2023) Effects of plant diversity on species-specific herbivory: patterns and mechanisms. Oecologia 201:1053–1066

Bröcher M, Ebeling A, Bassi L, Medina-van Berkum P, van Dam NM, Eisenhauer N, Madaj AM, Unsicker S, Weigelt A, Meyer ST (2025) Plant–herbivore interactions depend on plant richness, plant and soil history. Funct Ecol 39:3672–3687

Castagneyrol B, Jactel H, Vacher C, Brockerhoff EG, Koricheva J (2014) Effects of plant phylogenetic diversity on herbivory depend on herbivore specialization. J Appl Ecol 51:134–141

Castagneyrol B, Jactel H, Moreira X (2018) Anti-herbivore defences and insect herbivory: interactive effects of drought and neighbours. J Ecol 106:2043–2057

Duffy JE, Cardinale BJ, France KE, McIntyre PB, Thébault E, Loreau M (2007) The functional role of biodiversity in ecosystems: incorporating trophic complexity. Ecol Lett 10:522–538

Fornoff F, Klein AM, Blüthgen N, Staab M (2019) Tree diversity increases robustness of multitrophic interactions. Proc R Soc B 286:20182399

Hambäck PA, Inouye BD, Andersson P, Underwood N (2014) Effects of plant neighborhoods on plant-herbivore interactions: resource dilution and associational effects. Ecology 95:1370–1383

Hector A, Hautier Y, Saner P, Wacker L, Bagchi R, Joshi J, Scherer-Lorenzen M, Spehn E, Bazeley-White E, Weilenmann M, Caldeira M, Dimitrakopoulos P, Finn J, Huss-Danell K, Jumpponen A, Mulder C, Palmborg C, Pereira J, Siamantziouras A, Terry A, Troumbis A, Schmid B, Loreau M (2010) General stabilizing effects of plant diversity on grassland productivity at multiple sites through population asynchrony and overyielding. Ecology 91:2213–2220

Jactel H, Moreira X, Castagneyrol B (2021) Tree diversity and forest resistance to insect pests: patterns, mechanisms and prospects. Annu Rev Entomol 66:277–296

Kneitel J, Chase JM (2004) Disturbance, predator, and resource interactions alter container community composition. Ecol Lett 85:2088–2093
Lee SXT, Amir Z, Moore JH, Gaynor KM, Luskin MS (2024) Effects of human disturbances on wildlife behaviour and consequences for predator-prey overlap in Southeast Asia. Nat Commun 15:1521
Letourneau DK, Armbrecht I, Salguero-Rivera B, Montoya-Lerma J, Jiménez-Carmona E, Constanza-Daza M, Escobar S, Galindo V, Gutiérrez C, Duque-López S, López-Mejía J, Acosta-Rangel AM, Herrera-Rangel J, Rivera L, Saavedra CA, Torres AM, Reyes-Trujillo A (2011) Does plant diversity benefit agroecosystems? A synthetic review. Ecol Appl 21:9–21
Li Y, Schmid B, Schuldt A, Li S, Wang M-Q, Fornoff F, Staab M, Guo P-F, Anttonen P, Chesters D, Bruelheide H, Zhu C-D, Ma K, Liu X (2023) Multitrophic arthropod diversity mediates tree diversity effects on primary productivity. Nat Ecol Evol 7:832–840
Moreira X, Abdala-Roberts L, Rasmann S, Castagneyrol B, Mooney KA (2016) Plant diversity effects on insect herbivores and their natural enemies: current thinking, recent findings, and future directions. Curr Opin Insect Sci 14:1–7
Ruttan A, Lortie CJ (2015) A systematic review of the attractant-decoy and repellent-plant hypotheses: do plants with heterospecific neighbours escape herbivory? J Plant Ecol 8:337–346
Scherber C, Eisenhauer N, Weisser WW, Schmid B, Voigt W, Fischer M, Schulze ED, Roscher C, Weigelt A, Allan E, Bessler H, Bonkowski M, Buchmann N, Buscot F, Clement LW, Ebeling A, Engels C, Halle S, Kertscher I, Klein AM, Koller R, Konig S, Kowalski E, Kummer V, Kuu A, Lange M, Lauterbach D, Middelhoff C, Migunova VD, Milcu A, Muller R, Partsch S, Petermann JS, Renker C, Rottstock T, Sabais A, Scheu S, Schumacher J, Temperton VM, Tscharntke T (2010) Bottom-up effects of plant diversity on multitrophic interactions in a biodiversity experiment. Nature 468:553–556
Schuldt A, Ebeling A, Kunz M, Staab M, Guimarães-Steinicke C, Bachmann D, Buchmann N, Durka W, Fichtner A, Fornoff F, Härdtle W, Hertzog L, Klein AM, Roscher C, Schaller J, von Oheimb G, Weigelt A, Weisser W, Wirth C, Zhang J, Bruelheide H, Eisenhauer N (2019) Multiple plant diversity components drive consumer communities across ecosystems. Nat Commun 10:1460
Staab M, Schuldt A (2020) The influence of tree diversity on natural enemies – a review of the "enemies" hypothesis in forests. Curr For Rep 6:243–259
Staab M, Liu X, Assmann T, Bruelheide H, Buscot F, Durka W, Erfmeier A, Klein A-M, Ma K, Michalski S, Wubet T, Schmid B, Schuldt A (2021) Tree phylogenetic diversity structures multitrophic communities. Funct Ecol 35:521–534
Staab M, Pietsch S, Yan H, Blüthgen N, Cheng A, Li Y, Zhang N, Ma K, Liu X (2023) Dear neighbor: trees with extrafloral nectaries facilitate defense and growth of adjacent undefended trees. Ecology 104:e4057
Tilman D, Wedin D, Knops J (1996) Productivity and sustainability influenced by biodiversity in grassland ecosystems. Nature 379:718–720
Underwood N, Inouye BD, Hambäck PA (2014) A conceptual framework for associational effects: when do neighbors matter and how would we know? Q Rev Biol 89:1–19
Vázquez-González C, Castagneyrol B, Muiruri EW, Barbaro L, Abdala-Roberts L, Barsoum N, Fründ J, Glynn C, Jactel H, McShea WJ, Mereu S, Mooney KA, Morillas L, Nell C, Nock CA, Paquette A, Parker JD, Parker WC, Roales J, Scherer-Lorenzen M, Schuldt A, Verheyen K, Weih M, Yang B, Koricheva J (2024) Tree diversity enhances predation by birds but not by arthropods across climate gradients. Ecol Lett 27:e14427
White JA, Whitham TG (2000) Associational susceptibility of cottonwood to a box elder herbivore. Ecology 81:1795–1803

Role of Tree Traits in Mediating Effects of Forest Diversity on Herbivores

2

Juri A. Felix and Julia Koricheva

Abstract

Plant traits, including defensive, nutritional, morphological and physiological traits, play an important role in shaping plant-herbivore interactions. In plant communities composed of multiple species, herbivores may be influenced not just by the traits of the plant they are feeding on, but also by traits of neighbouring plants which might attract or repel herbivores. In addition, the presence of heterospecific neighbours can influence the trait expression of the focal plants themselves, which can in turn affect their resistance to herbivory. In this chapter, we review evidence for both types of these overlooked mechanisms of associational effects in forests. Many studies have shown that non-host volatiles and the presence of spiny or thorny neighbours might 'hide' focal plants from herbivores in mixed stands, leading to associational resistance. The same traits of neighbouring plants may also affect natural enemies of herbivores, either enhancing or weakening indirect plant defences, although these effects have been less researched. Results from studies assessing diversity-mediated intraspecific variation in trait expression have been highly variable, with examined traits differing in their direction and magnitude of response to diversity. This variability in focal tree trait responses is driven by species-specific effects of neighbours on abiotic and biotic conditions, as well as resource availability for the focal trees. Overall, tree diversity-mediated inter- and intraspecific trait variation can contribute to either associational resistance or associational susceptibility, depending on context. Relatively few studies that have documented changes in traits of focal trees in mixed species stands have established direct links with herbivory. We discuss why trait variation often serves as a poor predictor of herbivory and highlight

J. A. Felix (✉)
Department of Biological Sciences, Royal Holloway University of London, Egham, UK

Royal Botanic Gardens, Kew, UK
e-mail: Juri.Felix.2020@live.rhul.ac.uk

J. Koricheva
Department of Biological Sciences, Royal Holloway University of London, Egham, UK

X. Moreira et al. (eds.), *Effects of Plant Diversity on Plant-Herbivore Interactions*, Ecological Studies 255, https://doi.org/10.1007/978-3-032-26461-9_2

several research biases, including a disproportionate focus on leaf traits and chewing insect herbivores while other plant-herbivore interactions remain understudied.

2.1 Introduction

Plant traits play an important role in mediating plant-herbivore interactions (Awmack and Leather 2002; Ehrlich and Raven 1964; Stamp 2003). Defensive traits such as bark, resins, waxes, and spines, as well as secondary metabolites including tannins, flavonoids, terpenes, and alkaloids, help protect plants from herbivory. Likewise, indirect defences such as volatile organic compounds (VOCs) and extrafloral nectaries can reduce herbivory by recruiting natural enemies (Farmer 2014; Moreira et al. 2016). Many plants also deploy induced defences in response to volatile cues, herbivory, or other forms of tissue damage (e.g. by increasing defence levels or producing new chemical compounds following herbivory) (Eyles et al. 2010; Vázquez-González et al. 2022).

In contrast to the above defensive traits, some plant nutritional traits may enhance plant quality and lead to higher herbivory. For instance, the performance of many insect herbivores is limited by nitrogen; therefore, higher nitrogen content in plants is expected to increase palatability and overall herbivory (Mattson 1980; Zvereva et al. 2024). Similarly, higher concentrations of phosphorus and sugars are associated with improved plant quality to herbivores (Awmack and Leather 2002; Le Gall and Behmer 2014; Rode et al. 2017). High carbon content, on the other hand, tends to lower plant quality as it negatively correlates with nitrogen content and is linked to higher concentrations of carbon-based phenolic compounds (Schädler et al. 2003). Lastly, plant morphological or structural traits, such as a decrease in specific leaf area (SLA) and increase in leaf mass per area (LMA) or leaf thickness, may also reduce plant quality by limiting tissue digestibility (Zhu et al. 2024).

Initial research on plant antiherbivore mechanisms considered herbivory as a pairwise interaction between the herbivore and its host plant (i.e. a plant within the feeding range of a given herbivore) and was solely focused on traits of the host plant species (Zamora et al. 1999). This included seminal work by Feeny (1970), who linked springtime feeding by Lepidoptera larvae on oak leaves to leaf traits such as toughness, tannins, and nitrogen content.

However, ecological processes such as herbivory are known to depend on a wider ecological context including the species composition of the surrounding plant community (i.e. neighbouring plants). These community composition effects on focal plants (i.e. the plants on which herbivory is being measured) can also be linked to plant traits in two main ways (Fig. 2.1):

(a) Interspecific trait effects: herbivory on focal plants may be influenced not only by their own traits but also by the traits of neighbouring plants, which might attract or repel herbivores (associational effects) (Barbosa et al. 2009).

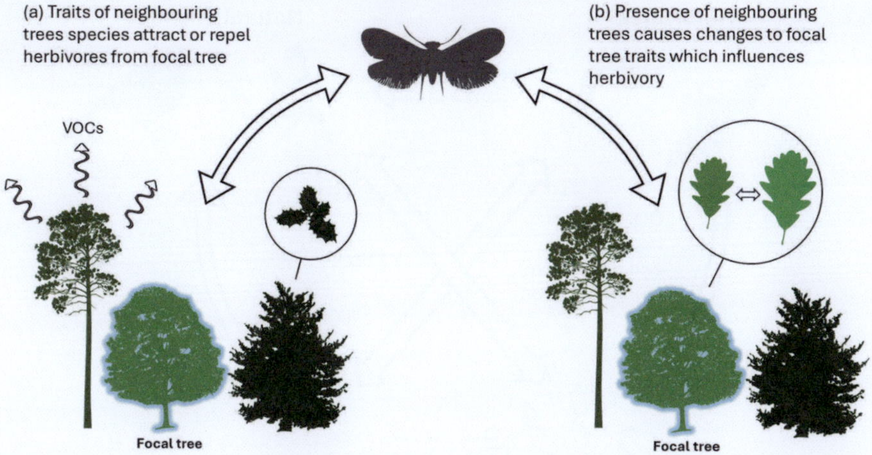

Fig. 2.1 Heterospecific neighbours may influence herbivory on focal trees either by (**a**) expressing traits that attract or repel herbivores, thereby altering herbivore abundance and damage; or (**b**) modifying focal tree traits that affect plant quality, which in turn can influence herbivory

(b) Intraspecific trait variation: the traits of focal plants may change due to the influence of neighbouring plants through competitive or facilitative mechanisms which could in turn affect plant resistance to herbivory (Felix et al. 2023).

In this chapter, we review the evidence for both of these mechanisms in forests, focusing first on the influence of neighbouring tree traits on herbivores, and then addressing effects of neighbourhood diversity on intraspecific trait variation in focal trees. While these two mechanisms likely interact to shape herbivore feeding patterns, their joint effects have rarely been examined, so we assess them separately in this chapter. Our emphasis is placed on the latter mechanism, reflecting the greater body of research available on intraspecific trait variation.

2.2 Interspecific Trait Effects

2.2.1 Effects of Traits of Neighbouring Species on Focal Tree Resistance to Herbivory

Traits of neighbouring plants can influence tree resistance to herbivores through both direct effects on herbivores and via effects on natural enemies. We will first explore direct effects of neighbouring plant traits on herbivores and then the indirect effects via natural enemies.

Due to their ability to travel long distances in the environment, plant volatiles have been commonly studied as mediators of plant-plant, plant-herbivore and multitrophic interactions (Schuman 2023). While specialist herbivores are attracted to their host's volatiles, the presence of non-host volatiles can have an inhibitory effect

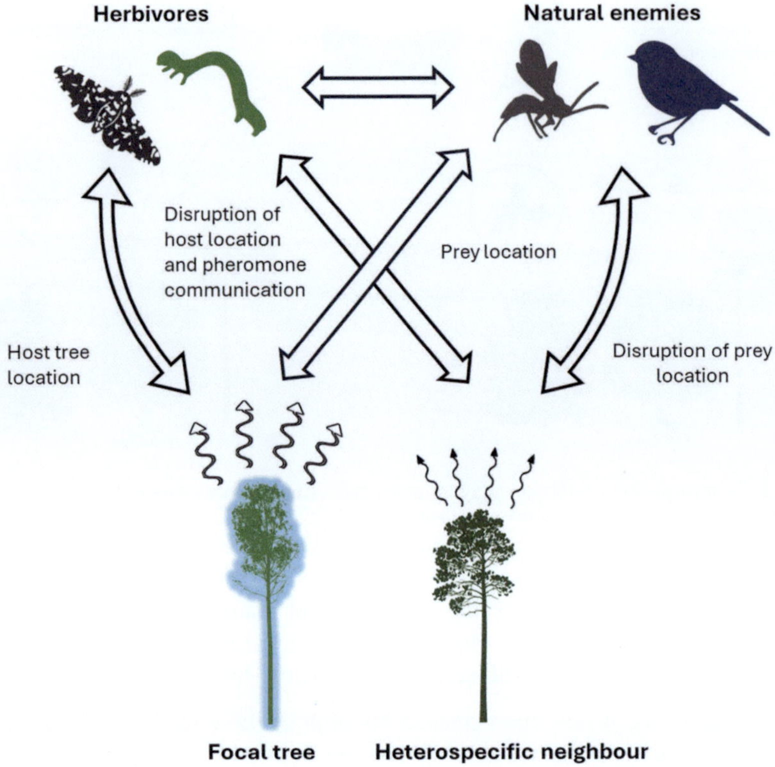

Fig. 2.2 Mechanisms through which volatile organic compounds (VOCs) emitted by the focal trees and neighbouring trees may affect the resistance of focal trees to herbivores

(Fig. 2.2). Such interference with host selection is a key mechanism underlying associational resistance. The semiochemical diversity hypothesis (Zhang and Schlyter 2003) predicts that mixed-species plant stands with higher semiochemical diversity can disturb olfactory-guided host choice, thereby reducing the risk of specialist herbivore outbreaks. This hypothesis is supported by studies showing that volatiles from non-host angiosperm trees can chemically mask neighbouring plants and deter specialist herbivores of conifers, such as bark beetles (Zhang and Schlyter 2004) and the pine processionary moth (Jactel et al. 2011). Non-host plant volatiles can also interfere with pheromone communication in herbivores by creating a 'noisy' chemical background that reduces the signal-to-noise ratio. For instance, Wang et al. (2016) demonstrated that non-host terpenoids released by conifers and *Eucalyptus* trees negatively affect both male orientation to sex pheromone and female calling behaviour (pheromone emission) in the specialist herbivore *Plutella xylostella*, potentially lowering mating success. Similarly, non-host bark volatiles have been shown to significantly reduce female attraction to aggregation pheromones in the bark beetle *Ips typographus* (Zhang 2003). It is not only the unique volatiles from non-host plants that can interfere with host selection by specialist herbivores. Many VOCs, particularly terpenoids, are common to phylogenetically distant plant species (Vivaldo et al. 2017). Zu et al. (2020) propose that these seemingly redundant compounds evolved as part of an informational 'arms race' between plants and herbivores, potentially making it more difficult for specialist herbivores to locate their host plants in species-rich communities. Zu et al. (2020) also provide field data on plant VOC profiles and plant-herbivore interactions from a tropical dry forest to support their theory. In addition to volatile compounds, other traits of neighbouring plants which contribute to their low palatability (e.g. physical and chemical defences as well as low levels of nutrients) can protect focal plants from herbivory, particularly by generalist mammalian herbivores. For instance, the presence of thorny or spiny neighbours can create safe sites for the establishment of palatable tree species (Bakker et al. 2004; Smit et al. 2006). Such associational refuges are particularly important during the seedling and sapling stages of focal trees as these stages are particularly vulnerable to mammalian herbivory.

The same traits of neighbouring plants which affect herbivores of focal plants can also influence natural enemies of these herbivores and hence the effectiveness of indirect defences. For instance, non-host volatiles may interfere with prey location by natural enemies of herbivores (Fig. 2.2, Monteith 1960; Zhang and Schlyter 2010), which could in turn disrupt the top-down control of herbivores in mixed-species forests. Zhang and Schlyter (2010) proposed that natural enemies of herbivores are able to recognise the volatiles of non-hosts of their prey and use them as signals to avoid unsuitable habitats. To test this hypothesis, they studied the behaviour of the predatory beetle *Thanasimus formicarius* in relation to angiosperm plant volatiles, which are non-host volatiles for its conifer-feeding bark beetle prey, and found that the presence of non-host volatiles inhibited predator attraction to prey signals both in the lab and in the field. Consequently, the signals used by natural enemies to locate prey can be masked by the chemical background of non-host neighbours in mixed species stands.

On the other hand, indirect defences of neighbouring plants may also benefit focal plants. For instance, a study in the Brazilian Cerrado (Moura and Del-Claro 2023) demonstrated that the presence of plants with extra floral nectaries indirectly benefited the liana *Smilax polyantha* by attracting mutualistic ant species, resulting in a threefold decrease in herbivory on *S. polyantha*. Moura and Del-Claro (2023) also showed that the effect was observed in the other direction as well; i.e. *S. polyantha* with artificial nectaries had significantly increased ant visitation on neighbouring plants. Another example of how indirect defences in neighbouring plants can provide protection from herbivory is provided by Riihimäki et al. (2005). Their study, conducted in a forest diversity experiment in Finland, revealed that wood ants forage more on birch trees in birch-pine mixtures than in birch monocultures, because pine trees hosted colonies of ant-tended aphids. This increased ant activity resulted in lower survival of autumnal moth (*Epirrita autumnata*) larvae and pupae on birch trees in pine-birch mixtures as compared to birch monocultures.

Overall, effects of neighbouring plant traits can lead to both associational resistance and associational susceptibility of focal trees to herbivores depending on the direction and relative importance of trait effects on insect herbivores and the natural enemies of herbivores.

2.3 Intraspecific Trait Variation

2.3.1 Neighbour-Induced Changes in Focal Tree Traits and Their Impact on Herbivory

While interspecific differences are the dominant source of trait variation in species-rich forests (Fig. 2.1a), reflecting the distinct evolutionary histories and defence strategies of different species (Wright et al. 2004), a substantial proportion of variation in traits relevant to herbivory also occurs within species (Fig. 2.1b). A meta-analysis by Siefert et al. (2015) found that intraspecific variation, whereby individuals of the same species express different traits, can account for up to 25% of total trait variation within plant communities. Among other factors, this variation is linked to local growing conditions and plant-plant interactions (both facilitative and competitive), which are a function of neighbouring species composition (Underwood et al. 2014; Westerband et al. 2021). This represents an additional pathway through which associational effects can emerge in mixed forests (Felix et al. 2023). In the following section, we explore this in detail, focusing on the mechanisms by which plant neighbourhood diversity and composition influences trait expression in focal trees and its effects on herbivores and herbivory.

2.3.2 Neighbourhood-Mediated Trait Variation in Tree Species Mixtures

An increasing number of studies in recent years have assessed the effects of tree diversity-mediated intraspecific trait variation as a driver of associational effects and other ecosystem functions (Massey et al. 2006; Muiruri et al. 2019; Poeydebat et al. 2020; Proß et al. 2021; Rosado-Sánchez et al. 2018a; Wang et al. 2024). Although herbivores feed on all plant parts and tissues, most research to date has focused on leaf traits and the damage caused by insect leaf chewers (Jactel et al. 2021), likely due to the relative ease of measuring this type of damage compared to that caused by other feeding guilds, such as stem borers or phloem feeders. Findings have been variable, both among and within individual studies, with increasing forest diversity reported to either increase or decrease the expression of traits influencing plant quality, depending on the context (Felix et al. 2023). Moreover, few studies have demonstrated causal relationships between neighbourhood-mediated variation in defensive traits and herbivory (i.e. whether neighbourhood-mediated trait variation alters herbivory on focal plants). Instead, most have found either no effects of measured traits on herbivory or that the effects of defensive traits on herbivory were independent of tree diversity effects (Table 2.1). Much of this research relies on long-term tree diversity experiments such as those in TreeDivNet (Depauw et al.

2024), which include replicated monocultures and species mixture plots, in some cases further accounting for species composition within mixtures. Even within these highly controlled experiments, the role of tree defences in mediating associational effects has been inconsistent.

This inconsistency can be clearly seen by comparing the results of several recent studies that used TreeDivNet experiments to explore the links between diversity, leaf traits and herbivory using oaks (*Quercus* ssp.) as focal species. Castagneyrol et al. (2017) measured physical leaf traits of English oaks (*Q. robur*) growing in monocultures and in various species-mixture plots in the ORPHEE forest diversity experiment in France. They found that neighbourhood effects differed between individual traits and depended on the species composition of each plot. Leaf toughness did not differ significantly between oak monocultures and oak-birch plots but was significantly lower in oak-pine and oak-pine-birch plots, whereas leaf thickness was significantly lower in all species-mixture plots compared to monocultures. Neither leaf thickness nor leaf toughness correlated with variation in herbivory. In contrast, we recently found that higher leaf thickness in *Q. robur* growing in the Climate Match tree diversity experiment in England was associated with lower leaf herbivory. However, this relationship was observed only for one of the two oak provenances examined and was independent of stand diversity (Felix et al. 2025). Furthermore, Galmán et al. (2022) studied the herbivory and leaf traits of holm oaks (*Q. ilex*) in the ORPHEE tree diversity experiment and found no effect on herbivory, despite both phenolics and VOCs being reduced in species mixtures containing pines. The authors speculate that this was because the oaks were growing among taller pine trees in species mixture plots; shading from tall neighbours led to reduced phenolic compounds in the oak leaves but also reduced the apparency of oaks to herbivores; thus the associational effect was neutral overall.

2.4 Why Is Diversity-Mediated Trait Expression Inconsistent?

We recently conducted a meta-analysis synthesising the effects of plant species diversity on eight leaf traits that can influence herbivory, incorporating data from many of the studies presented in Table 2.1 (Felix et al. 2023). Our analysis revealed that the mean effect of stand diversity on most traits was non-significant and highly variable, with traits often exhibiting large increases or decreases between monocultures and species mixtures. We also evaluated the absolute magnitude of trait responses to quantify the degree of trait plasticity irrespective of the direction of change. Under this approach, seven out of eight examined traits (SLA, LDMC, N, C, phenolics, terpenes, and thickness) showed significant responses to species diversity. These findings suggest that the effects of plant diversity on leaf traits can be strong but are highly context-dependant, resulting in either associational resistance or associational susceptibility, or null effects on herbivory if different trait responses counteract one another.

The variable effects of tree diversity on traits and defences can be better understood by examining the mechanisms through which diversity influences trait

Table 2.1 Summary of studies assessing how tree diversity affects intraspecific trait variation associated with tree quality for herbivores

Study	Focal tree species	Diversity treatments	Traits measured	Effect of neighbourhood diversity on traits	Did diversity-mediated trait variation influence herbivory?
Massey et al. (2006)	*Shorea leprosula*	Monocultures vs. five-species mixtures plots	**Leaf** *Phenolics, toughness*	Phenolics higher in species mixtures for young leaves but lower for old leaves	No Herbivory was lower in species mixtures but not due to higher defences
Moreira et al. (2012)	Pines (*P. pinaster*, *P. radiata*, *P. sylvestris*)	Monocultures vs. two- and three-species mixture plots	***Stem*** *Phenolics, non-volatile resin, nitrogen, antioxidant capacity, nonstructural carbohydrates*	No effect	Herbivory not measured
Moreira et al. (2014)	Large-leaf mahogany (*Swietenia macrophylla*)	Monocultures vs. four-species mixtures plots	**Leaf** *Phenolics* **Stem** *Phenolics*	Species and genotypic diversity increased phenolics in stems but not in leaves	Herbivory not measured
Forey et al. (2016)	European beech (*Fagus sylvatica*)	Beech forest stands classified as monocultures or as species mixtures	**Leaf** *Phenolics, LDMC, LMA*	Phenolics, LDMC, and LMA reduced in species mixtures, with larger reductions seen at higher species richness	Herbivory not measured
Castagneyrol et al. (2017)	English oak (*Quercus robur*)	Monocultures vs. two- and three-species mixture plots	**Leaf** *Toughness, thickness, SLA, LDMC, C:N ratio*	Toughness and thickness lower and SLA higher in species mixtures but only significant for certain plot types	No Herbivory was negatively associated with C:N ratios, but this was independent of diversity effects
Moreira et al. (2017)	English oak (*Quercus robur*)	Field sites with high and low density of oaks	**Leaf** *Phenolics, SLA, water content*	No	No

Castagneyrol et al. (2018)	Silver birch (*Betula pendula*)	Monocultures vs. two- and three-species mixture plots	**Leaf** *Phenolics*, *C:N ratio*, *water content*, *sugars*, *starch*	Total phenolics reduced in mixed stands, other traits not affected	No Trees growing in species mixtures had lower levels of leaf phenolics and experienced associational susceptibility; however, leaf chemistry had low explanatory power
Rosado-Sánchez et al. (2018a)	Ziricote (*Cordia dodecandra*)	Monocultures vs. four-species mixtures plots	**Leaf** *Phenolics*, *trichomes*, *SLA*	Marginal reduction of trichome density in species mixtures, SLA significantly reduced in species mixtures	No A positive correlation between phenolics and beetle damage was found, but it was independent of diversity effects
Rosado-Sánchez et al. (2018b)	Six tropical tree species	Monocultures vs. four-species mixtures plots	**Leaf** *Phenolics*	Lower levels of phenolics in monocultures overall and in two of six species when examined seperately	Herbivory not measured
Muiruri et al. (2019)	Silver birch (*Betula pendula*)	Monocultures vs. 2-, 3-, and 5-species mixture plots	**Leaf** *Phenolics*, *toughness*, *thickness*, *SLA*, *LDMC*, *leaf area*, *C*, *N*, *water content*, *acid detergent fibre*	No effect of species richness on leaf traits, however leaf area was significantly lower in stands with higher proportions of non-host trees (e.g. host dilution)	Partially Host dilution-associated decreases of leaf area correlate with lower herbivory Other traits influenced herbivory but independently from diversity effects

(continued)

Table 2.1 (continued)

Study	Focal tree species	Diversity treatments	Traits measured	Effect of neighbourhood diversity on traits	Did diversity-mediated trait variation influence herbivory?
Bellone et al. (2020)	Scots pine (*Pinus sylvestris*)	Boreal forest stands with high and low density of pine	**Leaf** *Resin acids* *Nitrogen*	No plot-level effects but nitrogen content was lower in neighbourhoods with high proportions of pine	Herbivory not measured but sawfly larvae survival higher when nitrogen content of pine needles decreased in neighbourhoods with high proportion of pines
Ferlian et al. (2020)	Ten European angiosperm tree species	Monocultures vs. two-species mixture plots	**Leaf** *Phenolics, amino acids, sugars, C, N, P*	No effect	No
Poeydebat et al. (2020)	Silver birch (*Betula pendula*)	Monocultures vs. species mixture plots across 12 tree diversity experiments in Europe and North America	**Leaf** *Phenolics*	Phenolics higher in species mixtures	Higher phenolics in species mixtures associated with reduced herbivory but only in experiments at more northern latitudes
Proß et al. (2021)	Five tropical tree species	Monocultures vs. three- and six-species mixture plots	**Leaf** *SLA, LDMC, C, N, lignin, cellulose, hemicellulose*	N and hemicellulose decreased in plots with higher species richness for *Anacardium excelsum* but not other species	Herbivory not measured

Galmán et al. (2022)	Holm oak (*Quercus ilex*)	Monocultures vs. two- and three-species mixture plots	**Leaf** *Phenolics, VOCs*	Flavan-3-ols higher and VOCs lower in plots containing pines	No
Jia et al. (2023)	Nine broadleaf species	Shannon diversity estimated within a 20 m radius around each focal tree	**Leaf** *SLA, thickness, C:N ratio, sodium, silica*	C:N ratio showed slight decrease as the phylogenetic diversity of neighbouring trees increased	No
Wang et al. (2024)	12 broadleaf species	Monocultures vs. 4-, 8-. 16-species mixture plots	**Leaf** *Leaf area, SLA, thickness, water content, N, C, P, C:N ratio, LDMC*	Neighbourhood tree species richness and host dilution both associated with higher SLA and C:N ratio and lower N and leaf thickness	Higher foliar C:N ratio in species mixtures correlated with associational resistance, higher P associated with lower herbivory but independent from diversity effects
Felix et al. (2025)	English oak (*Quercus robur*)	Monocultures vs. three-species mixture plots	**Leaf** *Phenolics, thickness, SLA*	Hydrolysable tannins higher in species mixture plots than in monoculture plots but for only one of the two oak provenances examined	No Higher leaf thickness associated with reduced herbivory but only for one of two oak provenances examined and independently from diversity effects
Felix et al. (2026)	Scots pine (*Pinus sylvestris*) and Norway spruce (*Picea abies*)	Monocultures vs. two- and three-species mixture plots	**Leaf** *Terpene content, dry needle mass*	Terpene content in pine higher in pine-birch mixtures but not in pine-spruce mixtures, while spruce dry needle mass lowest in the presence of conspecifics	Herbivory not measured

expression and how these mechanisms vary with the specific characteristics of different mixture types. These mechanisms are discussed in the following section, drawing on our meta-analysis and the broader ecological literature.

2.4.1 Pathways Leading to Diversity-Mediated Trait Variation

Depauw et al. (2024) outlined three major mechanisms by which tree species diversity can influence the performance of focal trees. We adapted this conceptual framework to illustrate the effects of neighbouring tree species diversity on focal tree traits (Fig. 2.3). These mechanisms include the following:

- Resource availability refers to availability and accessibility of resources essential for plant growth and functioning, including light, water, and nutrients. Resource-rich environments may promote the expression of more acquisitive traits (e.g. high SLA, low levels of chemical defences), whereas the opposite may occur in resource poor environments.
- Biotic growing conditions encompass the influence of aboveground and belowground communities, including fungi, microbes, herbivores, and soil biota. Antagonistic interactions with herbivores or pathogens can trigger the expression of induced defences, whereas mutualistic interactions with mycorrhizal fungi or N-fixing bacteria may increase nutrient uptake leading to increased foliar nitrogen.
- Abiotic growing conditions represent the physical and chemical environment in which plants grow, such as microclimate (temperature, humidity, wind) and soil

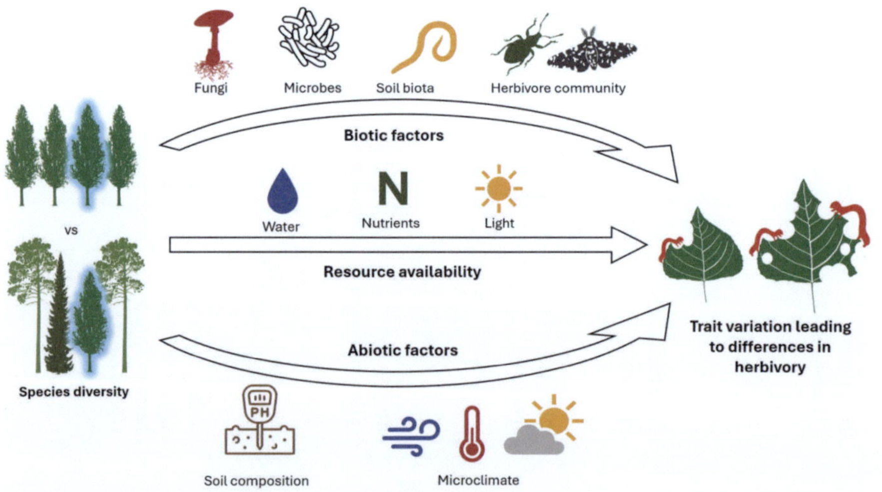

Fig. 2.3 Conceptual diagram illustrating how tree species diversity can influence focal plant traits and herbivory through changes in resource availability, biotic factors, and abiotic factors. Adapted from Felix et al. (2023) and Depauw et al. (2024)

properties (pH, texture, composition). These factors shape the wider ecological context in trees which are growing so may influence growth strategies and trait expression similarly to resource availability.

These mechanisms do not operate in isolation but instead interact in complex pathways, jointly influencing trait expression in trees. We discuss these pathways in more detail below.

2.4.2 Trait Variation Mediated by Resource Availability: Growth-Defence Trade-Offs and N-Fixers

Trees growing in monocultures are thought to experience stronger competition for water and nutrients than those in species mixtures, as all neighbouring trees occupy similar functional niches with comparable rooting depths, canopy structures, and resource requirements (Pretzsch 2014). This resource-limited environment can lead trees to adopt more conservative growth strategies and favour the expression of leaf traits that maximise longevity, such as low SLA and foliar nitrogen, and high LDMC and foliar carbon (Davrinche and Haider 2021; Reich 2014). Conversely, the presence of heterospecific neighbours in species mixtures may enhance resource availability through niche-partitioning or through facilitation (e.g. via nutrient-rich leaf litter) (Grossiord 2020; Wright et al. 2017). Trees with greater resource access may prioritise more acquisitive traits that promote rapid growth, such as high SLA and leaf nitrogen, which may coincide with reduced investment in physical and chemical defences (Davrinche and Haider 2021; Reich 2014). These relationships are sometimes framed as growth-defence trade-offs (Herms and Mattson 1992), although this concept is primarily applied to differences in life-history strategies between (Coley et al. 1985; Stamp 2003) and within plant species (Koricheva 2002). To date, it has only rarely been examined as a mechanism contributing to associational effects (Abdala-Roberts et al. 2014; Galmán et al. 2022; Rosado-Sánchez et al. 2018b).

Current research suggests that differences in abiotic conditions and resource availability between monocultures and mixed stands could modulate such trade-offs and that they are not driven solely by species richness but also depend on the presence of particular neighbouring species or plant functional groups (Trogisch et al. 2021). For instance, nitrogen-fixing plants (e.g. members of the Fabaceae family) are characterised by their ability to access atmospheric nitrogen (N) through symbiotic relationships with nitrogen-fixing bacteria in root nodules (Mathesius 2022). Although most benefits are retained by the N-fixing plant itself, studies have shown that these species can also enrich soil N, thereby increasing N availability for neighbouring plants and enhancing growth and biomass (Ortiz and Wolf 2024), although this effect is not universal (Staccone et al. 2021). In other words, N-fixers often act as natural fertilisers, influencing neighbouring tree growth and leaf traits in predictable ways.

The effects of plant trait diversity or composition can also extend to decomposition and nutrient cycling and potentially modulate trade-offs affecting plant traits.

The leaf litter of broadleaf trees decomposes more readily and contributes more organic matter to the soil than the waxy, lignin-rich needles of conifers (Hou et al. 2020), which can also improve soil fertility. For instance, boreal forests containing *Betula* ssp. have been shown to exhibit higher soil fertility compared to those dominated by conifers (e.g. *Pinus*, *Picea*) (Felton et al. 2016; Huuskonen et al. 2021). Consistent with some of the above mechanisms, our meta-analysis found that foliar nitrogen in non-N-fixing species was significantly higher in species mixtures containing N-fixing neighbours but not in diverse neighbourhoods without them. We also found that the presence of N-fixing neighbours was associated with lower concentrations of phenolic compounds in non-N-fixing focal plants, aligning with previous studies showing negative effects of fertilisation on leaf phenolics (Koricheva et al. 1998; Nybakken et al. 2018). This pattern is hypothesised to result from increased growth, which diverts carbon and other resources away from the production of carbon-based defences, i.e. diversity effects mediated by growth-defence trade-offs, although the exact mechanism is still debated (Hamilton et al. 2001).

2.4.3 Trait Variation Mediated by Abiotic Growing Conditions: Light Availability and Microclimate Buffering

Many leaf traits respond strongly to variation in light availability (Chen et al. 2020; Rozendaal et al. 2006), which may differ between species mixtures and monocultures due to canopy stratification and the presence of taller or shorter neighbours (Pretzsch 2014; Williams et al. 2020). Because plants rely on light for photosynthesis, leaves in shaded environments typically develop morphological traits that maximise light capture, such as higher specific leaf area (SLA) and lower thickness (Evans and Poorter 2001; Reich et al. 1997). Conversely, in high-light environments, the opposite trait shifts may occur: low SLA and increased thickness help reduce transpiration and mitigate UV damage (Gouveia and Freitas 2009; Verdaguer et al. 2012). Additionally, concentrations of several classes of secondary metabolites, many involved in leaf defences against herbivores, tend to increase with light availability (Ballaré 2014; Koricheva et al. 1998; Liu et al. 2023). This pattern may result from elevated metabolic rates or from the role of certain secondary metabolites in protecting against UV-induced oxidative stress (Roberts and Paul 2006).

In our meta-analysis (Felix et al. 2023), we found a positive correlation between plant species richness and SLA. High SLA has been linked to greater damage from chewing herbivores (Zhu et al. 2024) and is also associated with more acquisitive leaf traits, such as lower thickness, reduced toughness, and lower concentrations of chemical defences (Poorter et al. 2009; Reich 2014). We hypothesised that the positive effect of plant species richness on SLA is driven by shading, as trees in more diverse mixtures are more likely to grow alongside taller species that reduce light availability (Duarte et al. 2021). Light availability may also help explain another finding from our meta-analysis: phenolic compound concentrations were lower in mixtures of mature trees, but not in juvenile trees, but only when data for over-represented silver birch (*Betula pendula*) was excluded from models. We included tree life stage as a moderator in our analysis because we hypothesised that tree-tree interactions would exert stronger effects on leaf traits in more established woodlands that have undergone canopy closure (Jucker et al. 2020). Similar to the influence of species richness on SLA, shading effects may be more pronounced in mature woodlands than in juvenile stands. As a fast-growing pioneer species, silver birch often outcompetes heterospecific neighbours during early establishment (Huuskonen et al. 2021) and hence may experience greater light availability in mixtures than in monocultures—contrary to expected patterns. Indeed, much of the phenolics data in our meta-analysis came from a study by Poeydebat et al. (2020) which focused on silver birch in relatively young tree experiments and found that foliar phenolics were higher in species mixtures (Table 2.1). This likely explains why the significant reduction in phenolics in species mixtures only emerged in our meta-analysis model after silver birch data were removed. In addition to light availability, other abiotic conditions may differ between monocultures and mixtures. For instance, recent study by Schnabel et al. (2025) shows that tree species diversity increases forest temperature buffering through enhanced canopy density and structural diversity in species-rich stands. Resulting difference in microclimate in pure and mixed forest stands might affect plant traits, e.g. VOC emissions are known to be both light- and temperature-dependent (Dindorf et al. 2006), and hence future studies should explore the relationship between tree diversity, microclimate and tree traits.

2.4.4 Trait Variation Mediated by Biotic Growing Conditions: Induced Defences

Damage by herbivores can trigger induced defences in trees. VOCs such as green leaf volatiles, methyl salicylate, and methyl jasmonate are released by some plants following tissue damage and can induce the production of defences in other plant parts (Aljbory and Chen 2018; Kant et al. 2009). These signals can also trigger induced defences in neighbouring plants that 'eavesdrop' on volatile cues (Karban et al. 2004). This phenomenon is strongest between closely related plants from the same population but can also occur between heterospecific neighbours, albeit to a lesser extent (Karban et al. 2016; Karban and Shiojiri 2009).

It is therefore possible that trees growing in monocultures, species-poor mixtures, or species mixtures composed of closely related species may experience higher rates of defence induction in response to herbivory than those in more complex mixtures (Moreira and Abdala-Roberts 2019). For instance, Baier et al. (2002) found that Norway spruce trees in mixed stands with beech exhibited lower induced resin flow (a defensive response to bark beetle attack) compared to spruce trees in pure stands (although the authors did not test whether this was a result of reduced signalling via VOCs).

Neighbouring plants may also influence induced defences of focal plants by altering the availability of cues necessary for induction (e.g. physical damage or VOCs). For instance, Coverdale et al. (2018) reported that associational refuges provided by neighbouring spinescent *Acacia* trees in African savanna reduced induced responses in physically defended understory plants by lowering browsing damage. This example is a combination of both mechanisms illustrated in Fig. 2.1: defensive traits of neighbouring plants (spines) directly affected herbivory on focal plants and created associational refuge and in addition, neighbouring plants also affected defensive traits of focal plants through a relaxation of induced defences.

However, most research in this area has focused on herbaceous plants, and trees may often grow too far apart for volatile signals to effectively induce defences in neighbours (Eyles et al. 2010). Indeed, a recent study on Japanese beech (*Fagus crenata*) found that simulated leaf damage (by cutting leaves with scissors) triggered the emission of green leaf volatiles and reduced herbivore damage in nearby trees, but only within a 5–7 m radius of the damaged tree. This suggests that induced defence responses mediated by volatiles occur at highly localised scales (Hagiwara et al. 2021).

Beyond herbivory, trees also face significant threats from viral, bacterial, and fungal pathogens (Teshome et al. 2020) which can also induce tree defences. Leaf traits are known to play roles in defence against these additional biotic agents; for instance, tannins can be toxic to certain microorganisms (Field and Lettinga 1992), and many secondary metabolites exhibit antifungal properties (Zaynab et al. 2018). Similar to herbivory, pathogen infection and damage have been shown to vary with tree diversity (Nguyen et al. 2016; Rutten et al. 2021), thus creating different biotic environment for trees in monocultures and species mixtures. It is possible that trait variation exerts a stronger influence in mitigating these other biotic threats than it does in protecting trees from insect herbivores. More holistic studies addressing both pathogens and herbivory are therefore needed (e.g. Ferlian et al. 2020; Field et al. 2020) to better understand the role of biotic environment in trait variation and, vice versa, the role of trait variation in mediating multiple biotic pressures within diverse forests (Halliday et al. 2025).

2.5 Other Sources of Variation

2.5.1 Focal Tree Species Identity

The mechanisms described above are influenced by neighbouring species identity; however, their effects are rarely uniform and often depend on the focal species identity, as discussed in relation to silver birch above. For example, Rosado-Sánchez et al. (2018b) found that overall concentrations of phenolic compounds in the leaves of tropical tree species growing in mixed stands of the UADY tree diversity experiment were lower than in monocultures, but this pattern was significant for only two of the six species when analysed individually. Species-specific trait responses may depend on life history or functional strategies associated with growth rate, nutrient acquisition, and shade tolerance or other adaptations that influence the degree of trait plasticity under different competitive and environmental conditions (Bartkowicz and Paluch 2023; Bennett et al. 2016). Some studies report that light-demanding pioneer species exhibit greater trait variability in response to light availability than shade-tolerant species (Böhnke and Bruelheide 2013; Niinemets 2010; Tripathi et al. 2020). Our own recent findings further support this pattern, as we show that terpene concentrations in Scots pine, but not Norway spruce, respond to neighbourhood diversity (Felix et al. 2026). Results in other studies have been less consistent: Williams et al. (2020) found no effect of shade tolerance on diversity-driven trait plasticity but reported that angiosperms are more responsive than gymnosperms, whereas Proß et al. (2025) found mixed results, with SLA and LDMC showing greater responses to light availability in deciduous trees, but other traits, including leaf nitrogen and tannin content showing greater responses in evergreen trees.

Trait plasticity in response to diversity, competition, and abiotic factors can also vary within focal species (e.g. among geographic provenances or genotypes) (Cope et al. 2021; Felix et al. 2025; Wortemann et al. 2011). For instance, we recently examined the effects of stand diversity on leaf traits of oak (*Quercus robur*) in an assisted migration experiment in the United Kingdom and found that oaks of Italian provenance had significantly higher concentrations of hydrolysable tannins in species mixtures compared to monocultures, whereas oaks of local (English) provenance showed no significant difference between monocultures and mixtures (Felix et al. 2025).

2.5.2 Within-Individual Trait Variation

Within-individual (intra-individual) trait variation refers to differences in trait expression between different parts of the same tree, often driven by microclimatic heterogeneity within the canopy. It is well documented than sun-exposed leaves at the top of tree canopies typically exhibit lower SLA and higher thickness, toughness, and phenolics than leaves lower in the canopy, as these traits help protect against UV radiation and excessive transpiration (Proß et al. 2025; Scartazza et al. 2016). Recent studies by Sánchez-Bermejo et al. (2023, 2024) demonstrate that the degree of within-individual trait variation in focal trees can also respond to species richness, further complicating our understanding of diversity-mediated effects on leaf traits and herbivores.

2.5.3 Temporal Factors

Tree-tree interactions and the effects of different species compositions on abiotic factors are known to intensify over time as trees grow and diverge according to their individual life strategies (Jucker et al. 2016; Koricheva et al. 2025). The mechanisms underlying diversity-mediated trait variation discussed above may therefore only emerge at scales that have a measurable effect on leaf traits years or even decades after initial forest establishment. However, many experimental sites used to study the role of trait variation in driving associational effects are relatively young; for instance, the oldest experiment in TreeDivNet, the Satakuna experiment, was established in 1999 (Depauw et al. 2024). Consequently, the current evidence in this field may be inconsistent, because it relies on data that does not fully capture biodiversity-ecosystem functioning and plant-herbivore interactions typical of mature forests.

Moreover, as forests develop, diversity-mediated trait shifts may reverse due to changes in light availability and competitive dominance among species. For example, initial increases in leaf phenolics in pioneer species like silver birch as described by Poeydebat et al. (2020) may decline if these species are later outcompeted by taller, late-successional species.

2.5.4 Spatial Factors

Many ecological processes vary with the spatial scale at which they are measured, including relationships between diversity and ecosystem functioning (Chisholm et al. 2013; Thompson et al. 2018; Yue et al. 2005). For instance, Fan et al. (2020) found a positive relationship between tree diversity and tree growth in temperate forests of northeastern China at larger scales (diversity measured within radii over 9 m from focal trees), but a negative relationship at smaller scales (radii of 3–5 m). The authors proposed that diversity promotes both resource partitioning and facilitative effects, which generally benefit tree growth, and competition (e.g. due to canopy overlap), which tends to reduce growth. However, the negative effects of competition were most acute at highly localised scales where canopy overlap was greatest.

Similar mechanisms may explain the inconsistent effects of tree diversity on leaf traits in the current evidence base (Table 2.1), which primarily relies on data linked to species diversity at the plot or stand level rather than accounting for immediate neighbours. Higher canopy cover reduces light availability and induces predictable shifts in leaf traits, yet this effect is driven mainly by the nearest neighbours and may not consistently correlate with diversity at larger scales. Likewise, N-fixers or species producing nutrient-rich litter influence soil fertility most strongly for nearby trees rather than distant ones. Even in regularly planted plots, neighbour identity and influence can vary due to uneven growth and stochastic tree mortality (Fig. 2.4). Such scale-dependent effects may partly explain the high heterogeneity in our meta-analysis models, which encompassed a wide range of experimental setups, from pairwise comparisons of conspecific and heterospecific plants (e.g. Ormeño et al. 2007) to large plots in tree diversity experiments containing hundreds of individuals (e.g. Muiruri et al. 2019). Notably, recent research shows that species identity of the nearest neighbours predicts leaf traits more reliably than plot-level species diversity (Bellone et al. 2020; Davrinche and Haider 2021; Felix et al. 2026).

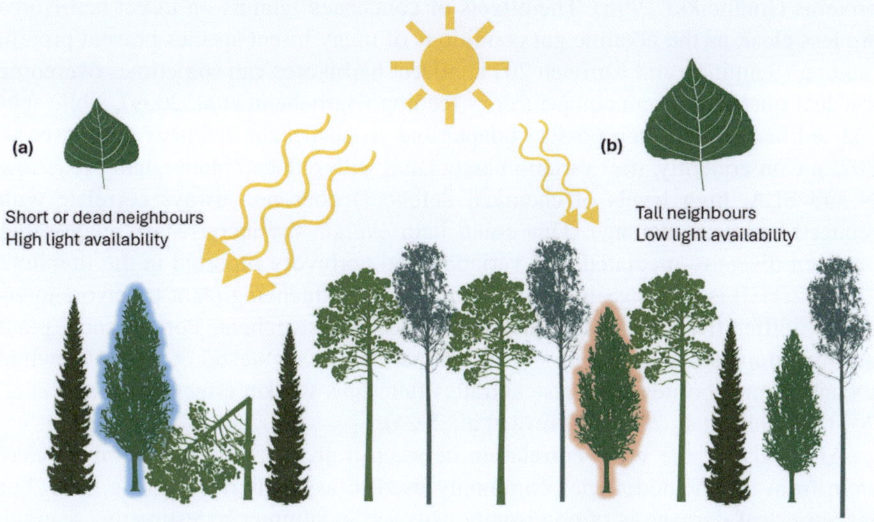

Fig. 2.4 Diagram illustrating how light availability varies with the nearest neighbours, influencing leaf traits of the focal tree: (**a**) In a four-species mixtures, the nearest neighbours of a focal tree (highlighted in blue) are either shorter, dead, or have canopy shapes that produce minimal shade. As a result, the focal tree experiences high light availability, which may lead to leaf adaptations that reduce transpiration and UV damage (e.g. lower SLA, increased thickness, and higher concentrations of phenolic compounds). (**b**) When the nearest neighbours are taller than a different focal tree (highlighted in orange), shaded conditions prevail, and the focal tree's leaf traits may shift in the opposite direction to (**a**) to maximise photosynthesis

2.6 Why Does Diversity-Mediated Intraspecific Trait Variation Have Inconsistent Associational Effects?

Having established that the strength and direction of tree diversity effects on leaf defences vary across a range of context-specific factors, an important question remains: why do these trait changes not consistently translate into predictable patterns of insect herbivory (Table 2.1)? Below, we explore several reasons why diversity-mediated variation in plant traits appears to be a weak predictor of herbivory levels.

2.6.1 Inconsistent Relationships Between Plant Traits and Measured Herbivory

Herbivore feeding patterns in response to plant quality are known to depend on taxa-specific adaptations (Farmer 2014). Condensed tannins are thought to reduce nutritional quality for mammalian herbivores by binding to proteins and reducing nitrogen availability (Barbehenn and Constabel 2011). However, some species, such as moose (*Alces alces*), counteract this by producing tannin-binding salivary proteins (Juntheikki 1996). The effects of condensed tannins on insect herbivores are less clear, as the alkaline gut conditions of many insect species prevent protein binding (Salminen and Karonen 2011). nIsect herbivores can sometimes overcome low leaf quality through compensatory feeding (Barbehenn et al. 2009), while specialised herbivores often possess adaptations to host plant defences (Leong et al. 2022). Consequently, trait variation associated with reduced plant quality (e.g. low N and SLA, high levels of chemical defences) does not always correlate with reduced herbivore damage. This could help explain the inconsistent relationship between diversity-mediated trait variation and herbivory reported in the literature (Table 2.1). It is also possible that the main traits influencing plant-herbivore interactions differ from those typically measured by researchers. For instance, plant nutritive traits are strong drivers of insect performance (Wetzel et al. 2016), while secondary metabolites and physical traits often show weaker effects (Carmona et al. 2011; Ruhnke et al. 2009; Zvereva et al. 2024).

Alternatively, the weak correlation between trait variation and herbivory may stem from the methodologies commonly used to assess herbivory (Fig. 2.5). For instance, leaf damage is often visually assessed in summer by estimating average chewing damage on a subset of leaves per tree (Muiruri et al. 2019). Although this approach captures cumulative damage, it does not indicate when the damage occurred or which taxa were responsible; factors critical for understanding the role of leaf traits in plant-herbivore interactions (Vanhakylä and Salminen 2023; Zvereva et al. 2024).

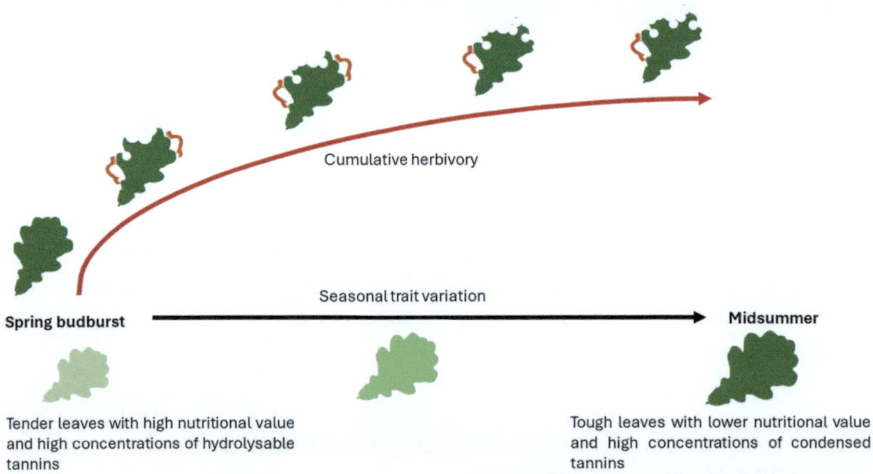

Fig. 2.5 Conceptual diagram illustrating seasonal variation in cumulative herbivory and leaf traits. Herbivory is typically most intense during the first weeks after leaf emergence (spring budburst), partly due to the high nutritional value of young leaves, and then accumulates more gradually through summer. As a counteradaptation, newly emerged leaves often contain higher concentrations of chemical defences (e.g. hydrolysable tannins in oaks). Leaf traits also shift over the growing season; for instance, oak leaves become tougher and accumulate condensed tannins as they age. Consequently, leaf traits measured in midsummer may not accurately reflect the leaf traits present during the peak herbivory in spring

Insect herbivores in temperate and boreal forests feed throughout the growing season, but peak damage often occurs after budburst in spring (Allison et al. 2023; Feeny 1970). Many herbivore species synchronise their emergence with their host plant's first flush of leaves, which are more tender and nutrient-rich early in development (Barton et al. 2019). Correspondingly, some trees may exhibit higher concentrations of certain secondary metabolites immediately after budburst (Salminen et al. 2004). As a result, a large proportion of cumulative herbivory recorded in midsummer likely occurred earlier, which could explain weak correlations in studies that only measured leaf traits once during midsummer.

The relationship between secondary metabolites and herbivory becomes even more complex when considering individual compounds, whose effects on herbivores can vary substantially (Moctezuma et al. 2014). Feeding trials using artificial diets allow for detailed assessment of individual molecules (Roslin and Salminen 2008). However, because the effects of secondary metabolites are highly taxa- and feeding guild-specific, this approach is impractical for trees that host hundreds of herbivorous species (Mitchell et al. 2022).

Future research on the link between leaf traits and herbivory should therefore include surveys of herbivore communities on focal trees (e.g. using the approach of Leong et al. (2024). These should target the period of peak herbivory after budburst to identify the herbivores responsible for most leaf damage. Once the most common or destructive species are identified, they can be collected for feeding trials to determine which leaf traits, if any, reliably predict herbivory (Roslin and Salminen 2008).

2.6.2 Overlooked Traits Driving Associational Effects

While commonly measured leaf traits related to leaf quality appear to play only a limited role in mediating associational effects, variation in other, often overlooked, plant traits may have stronger influences on plant-herbivore interactions in diverse forests. This is an area deserving further research, particularly as many forest pests target tissues other than leaves. For instance, bark beetles (*Dendroctonus* and *Ips* spp.) attack conifer stems, causing widespread destruction in temperate and boreal forests in the northern hemisphere (Fettig and Audley 2021; Gomez et al. 2023), while the peach tree borer (*Synanthedon exitiosa*) can severely damage the trunk and roots of cultivated stone fruit species (*Prunus* spp.) (Frank et al. 2020).

Moreira et al. (2014) found that both genotypic and species diversity were associated with significantly higher concentrations of phenolic compounds in the stems of big-leaf mahogany (*Swietenia macrophylla*), while leaf phenolics were unaffected by either treatment. This suggests that mechanisms driving trait expression may differ among plant parts, as has been shown for biomass allocation responses to environmental variables (Poorter et al. 2012). Leaves, being short-lived, may respond more strongly to seasonal changes in light and temperature, while roots and stems could be influenced by longer-term shifts such as soil nutrient availability or recurring drought.

Additionally, few studies have examined how indirect defences, such as VOCs and extrafloral nectaries respond to plant diversity (Moreira et al. 2016). Abiotic factors like light availability are known to affect VOC emissions (Holopainen and Gershenzon 2010; Pichersky and Dudareva 2020). For example, Galmán et al. (2022) reported lower VOC emissions in holm oak (*Quercus ilex*) when grown in oak-pine mixtures compared to oak monocultures or oak-birch mixtures. Similarly, nectar production by extrafloral nectaries in various species has been linked to light availability (Jones and Koptur 2015; Radhika et al. 2010). Thus, shading and other diversity-mediated mechanisms could alter VOC emissions and nectar availability, affecting the attraction of natural enemies such as parasitic wasps and ants, and consequently influencing levels of indirect defence (Bronstein et al. 2006; Heil 2008; Kessler and Heil 2011).

Finally, phenological events such as budburst and seasonal changes in trait expression also shape herbivory (Burdon and Bartlett 2022; Ju et al. 2017) and have been shown to vary in species-rich forests (Shen et al. 2024). The effect of phenology on herbivores may stem from seasonal variation of plant traits (e.g. Figure 2.5) but may also be linked to food availability and synchronicity with herbivore emergence (Barber and Fahey 2015; Sampaio et al. 2016). Future research should integrate these aspects to assess how diversity-mediated phenological variation influence plant-herbivore interactions.

References

Abdala-Roberts L, Moreira X, Cervera JC, Parra-Tabla V (2014) Light availability influences growth-defense trade-offs in big-leaf mahogany (*Swietenia macrophylla* king). Biotropica 46(5):591–597. https://doi.org/10.1111/btp.12133

Aljbory Z, Chen M-S (2018) Indirect plant defense against insect herbivores: a review. Insect Sci 25(1):2–23. https://doi.org/10.1111/1744-7917.12436

Allison JD, Paine TD, Slippers B, Wingfield MJ (eds) (2023) Forest entomology and pathology: volume 1: entomology. Springer International Publishing. https://doi.org/10.1007/978-3-031-11553-0

Awmack CS, Leather SR (2002) Host plant quality and fecundity in herbivorous insects. Annu Rev Entomol 47:817–844. https://doi.org/10.1146/annurev.ento.47.091201.145300

Baier P, Führer E, Kirisits T, Rosner S (2002) Defence reactions of Norway spruce against bark beetles and the associated fungus Ceratocystis polonica in secondary pure and mixed species stands. For Ecol Manag 159(1):73–86. https://doi.org/10.1016/S0378-1127(01)00711-3

Bakker ES, Olff H, Vandenberghe C, De Maeyer K, Smit R, Gleichman JM, Vera FWM (2004) Ecological anachronisms in the recruitment of temperate light-demanding tree species in wooded pastures. J Appl Ecol 41(3):571–582. https://doi.org/10.1111/j.0021-8901.2004.00908.x

Ballaré CL (2014) Light regulation of plant defense. Annu Rev Plant Biol 65(1):335–363. https://doi.org/10.1146/annurev-arplant-050213-040145

Barbehenn RV, Constabel CP (2011) Tannins in plant–herbivore interactions. Phytochemistry 72(13):1551–1565. https://doi.org/10.1016/j.phytochem.2011.01.040

Barbehenn RV, Jaros A, Lee G, Mozola C, Weir Q, Salminen J-P (2009) Hydrolyzable tannins as "quantitative defenses": limited impact against Lymantria dispar caterpillars on hybrid poplar. J Insect Physiol 55(4):297–304. https://doi.org/10.1016/j.jinsphys.2008.12.001

Barber NA, Fahey RT (2015) Consequences of phenology variation and oxidative defenses in Quercus. Chemoecology 25(5):261–270. https://doi.org/10.1007/s00049-015-0194-5
Barbosa P, Hines J, Kaplan I, Martinson H, Szczepaniec A, Szendrei Z (2009) Associational resistance and associational susceptibility: having right or wrong neighbors. Annu Rev Ecol Evol Syst 40(1):1–20. https://doi.org/10.1146/annurev.ecolsys.110308.120242
Bartkowicz L, Paluch J (2023) Morphological plasticity of six tree species with different light demands growing in multi-layered deciduous forests in Central Europe. Eur J For Res 142(5):1177–1195. https://doi.org/10.1007/s10342-023-01584-7
Barton KE, Edwards KF, Koricheva J (2019) Shifts in woody plant defence syndromes during leaf development. Funct Ecol 33(11):2095–2104. https://doi.org/10.1111/1365-2435.13435
Bellone D, Björkman C, Schmidt A, Gershenzon J, Klapwijk MJ (2020) Effect of forest stand type on host plant quality and direct and indirect effects on pine sawfly performance. Agric For Entomol 23(2):163–172. https://doi.org/10.1111/afe.12415
Bennett JA, Riibak K, Tamme R, Lewis RJ, Pärtel M (2016) The reciprocal relationship between competition and intraspecific trait variation. J Ecol 104(5):1410–1420. https://doi.org/10.1111/1365-2745.12614
Böhnke M, Bruelheide H (2013) How do evergreen and deciduous species respond to shade? Tolerance and plasticity of subtropical tree and shrub species of south-East China. Environ Exp Bot 87:179–190. https://doi.org/10.1016/j.envexpbot.2012.09.010
Bronstein JL, Alarcón R, Geber M (2006) The evolution of plant–insect mutualisms. New Phytol 172(3):412–428. https://doi.org/10.1111/j.1469-8137.2006.01864.x
Burdon RD, Bartlett MJ (2022) Putative biotic drivers of plant phenology: with special reference to pathogens and deciduousness. Ecol Evol 12(6):e8932. https://doi.org/10.1002/ece3.8932
Carmona D, Lajeunesse MJ, Johnson MTJ (2011) Plant traits that predict resistance to herbivores. Funct Ecol 25(2):358–367. https://doi.org/10.1111/j.1365-2435.2010.01794.x
Castagneyrol B, Bonal D, Damien M, Jactel H, Meredieu C, Muiruri EW, Barbaro L (2017) Bottom-up and top-down effects of tree species diversity on leaf insect herbivory. Ecol Evol 7(10):3520–3531. https://doi.org/10.1002/ece3.2950
Castagneyrol B, Jactel H, Moreira X (2018) Anti-herbivore defences and insect herbivory: interactive effects of drought and tree neighbours. J Ecol 106(5):2043–2057. https://doi.org/10.1111/1365-2745.12956
Castro Sánchez-Bermejo P (2024) Tree and mycorrhizal fungal diversity drive intraspecific and intraindividual trait variation in temperate forests: evidence from a tree diversity experiment. Funct Ecol. https://doi.org/10.1111/1365-2435.14549
Castro Sánchez-Bermejo P, Davrinche A, Matesanz S, Harpole WS, Haider S (2023) Within-individual leaf trait variation increases with phenotypic integration in a subtropical tree diversity experiment. New Phytol 240(4):1390–1404. https://doi.org/10.1111/nph.19250
Chen X, Sun J, Wang M, Lyu M, Niklas KJ, Michaletz ST, Zhong Q, Cheng D (2020) The leaf economics spectrum constrains phenotypic plasticity across a light gradient. Front Plant Sci 11
Chisholm RA, Muller-Landau HC, Abdul Rahman K, Bebber DP, Bin Y, Bohlman SA, Bourg NA, Brinks J, Bunyavejchewin S, Butt N, Cao H, Cao M, Cárdenas D, Chang L-W, Chiang J-M, Chuyong G, Condit R, Dattaraja HS, Davies S et al (2013) Scale-dependent relationships between tree species richness and ecosystem function in forests. J Ecol 101(5):1214–1224. https://doi.org/10.1111/1365-2745.12132
Coley PD, Bryant JP, Chapin FS (1985) Resource availability and plant antiherbivore defense. Science 230(4728):895–899. https://doi.org/10.1126/science.230.4728.895
Cope OL, Lindroth RL, Helm A, Keefover-Ring K, Kruger EL (2021) Trait plasticity and trade-offs shape intra-specific variation in competitive response in a foundation tree species. New Phytol 230(2):710–719. https://doi.org/10.1111/nph.17166
Coverdale TC, Goheen JR, Palmer TM, Pringle RM (2018) Good neighbors make good defenses: associational refuges reduce defense investment in African savanna plants. Ecology 99(8):1724–1736. https://doi.org/10.1002/ecy.2397
Davrinche A, Haider S (2021) Intra-specific leaf trait responses to species richness at two different local scales. Basic Appl Ecol 55:20–32. https://doi.org/10.1016/j.baae.2021.04.011

Depauw L, De Lombaerde E, Dhiedt E, Blondeel H, Abdala-Roberts L, Auge H, Barsoum N, Bauhus J, Chu C, Damtew A, Eisenhauer N, Fagundes MV, Ganade G, Gendreau-Berthiaume B, Godbold D, Gravel D, Guillemot J, Hajek P, Hector A et al (2024) Enhancing tree performance through species mixing: review of a quarter-century of TreeDivNet experiments reveals research gaps and practical insights. Curr For Rep. https://doi.org/10.1007/s40725-023-00208-y

Dindorf T, Kuhn U, Ganzeveld L, Schebeske G, Ciccioli P, Holzke C, Köble R, Seufert G, Kesselmeier J (2006) Significant light and temperature dependent monoterpene emissions from European beech (Fagus sylvatica L.) and their potential impact on the European volatile organic compound budget. J Geophys Res Atmos 111(D16). https://doi.org/10.1029/2005JD006751

Duarte MM et al (2021) High tree diversity enhances light interception in tropical forests. J Ecol 109(7):2597–2611. https://doi.org/10.1111/1365-2745.13669

Ehrlich PR, Raven PH (1964) Butterflies and plants: a study in coevolution. Evolution 18(4):586–608. https://doi.org/10.1111/j.1558-5646.1964.tb01674.x

Evans JR, Poorter H (2001) Photosynthetic acclimation of plants to growth irradiance: the relative importance of specific leaf area and nitrogen partitioning in maximizing carbon gain. Plant Cell Environ 24(8):755–767. https://doi.org/10.1046/j.1365-3040.2001.00724.x

Eyles A, Bonello P, Ganley R, Mohammed C (2010) Induced resistance to pests and pathogens in trees. New Phytol 185(4):893–908. https://doi.org/10.1111/j.1469-8137.2009.03127.x

Fan C, Tan L, Zhang C, Zhao X, Gao L, von Gadow K (2020) Scale-dependent effects of neighborhood biodiversity on individual tree productivity in a coniferous and broad-leaved mixed forest in China. Ecol Evol 10(15):8225–8234. https://doi.org/10.1002/ece3.6530

Farmer EE (2014) Leaf defence. Oxford University Press

Feeny P (1970) Seasonal changes in oak leaf tannins and nutrients as a cause of spring feeding by winter moth caterpillars. Ecology 51(4):565–581. https://doi.org/10.2307/1934037

Felix JA, Stevenson PC, Koricheva J (2023) Plant neighbourhood diversity effects on leaf traits: a meta-analysis. Funct Ecol 37(12):3150–3163. https://doi.org/10.1111/1365-2435.14441

Felix JA, Stevenson PC, Barsoum N, Koricheva J (2025) Stand diversity does not mitigate increased herbivory on climate-matched oaks in an assisted migration experiment. Plant Cell Environ 48(5):3620–3631. https://doi.org/10.1111/pce.15383

Felix JA, Stevenson PC, Koricheva J (2026) Species identity and spatial scale drive context-dependent tree diversity effects in a Finnish forest experiment. Oikos

Felton A, Nilsson U, Sonesson J, Felton AM, Roberge J-M, Ranius T, Ahlström M, Bergh J, Björkman C, Boberg J, Drössler L, Fahlvik N, Gong P, Holmström E, Keskitalo ECH, Klapwijk MJ, Laudon H, Lundmark T, Niklasson M et al (2016) Replacing monocultures with mixed-species stands: ecosystem service implications of two production forest alternatives in Sweden. Ambio 45(2):124–139. https://doi.org/10.1007/s13280-015-0749-2

Ferlian O, Lintzel E-M, Bruelheide H, Guerra CA, Heklau H, Jurburg S, Kühn P, Martinez-Medina A, Unsicker SB, Eisenhauer N, Schädler M (2020) Nutrient status not secondary metabolites drives herbivory and pathogen infestation across differently mycorrhized tree monocultures and mixtures. Basic Appl Ecol 55:110–123. https://doi.org/10.1016/j.baae.2020.09.009

Fettig CJ, Audley JP (2021) Conifer bark beetles. Curr Biol 31(9):R419–R420. https://doi.org/10.1016/j.cub.2021.03.039

Field JA, Lettinga G (1992) Toxicity of tannic compounds to microorganisms. In: Hemingway RW, Laks PE (eds) Plant polyphenols: synthesis, properties, significance. Springer US, pp 673–692. https://doi.org/10.1007/978-1-4615-3476-1_39

Field E, Castagneyrol B, Gibbs M, Jactel H, Barsoum N, Schönrogge K, Hector A (2020) Associational resistance to both insect and pathogen damage in mixed forests is modulated by tree neighbour identity and drought. J Ecol 108(4):1511–1522. https://doi.org/10.1111/1365-2745.13397

Forey E, Langlois E, Lapa G, Korboulewsky N, Robson TM, Aubert M (2016) Tree species richness induces strong intraspecific variability of beech (Fagus sylvatica) leaf traits and alleviates edaphic stress. Eur J For Res 135(4):707–717. https://doi.org/10.1007/s10342-016-0966-7

Frank DL, Starcher S, Chandran RS (2020) Comparison of mating disruption and insecticide application for control of peachtree borer and lesser peachtree borer (Lepidoptera: Sesiidae) in peach. Insects 11(10):658. https://doi.org/10.3390/insects11100658

Galmán A, Vázquez-González C, Röder G, Castagneyrol B (2022) Interactive effects of tree species composition and water availability on growth and direct and indirect defences in Quercus ilex. Oikos 2022(5):e09125. https://doi.org/10.1111/oik.09125

Gomez DF, Riggins JJ, Cognato AI (2023) Bark Beetles. In: Allison JD, Paine TD, Slippers B, Wingfield MJ (eds) Forest entomology and pathology. Springer International Publishing, pp 299–337. https://doi.org/10.1007/978-3-031-11553-0_10

Gouveia AC, Freitas H (2009) Modulation of leaf attributes and water use efficiency in *Quercus suber* along a rainfall gradient. Trees 23(2):267–275. https://doi.org/10.1007/s00468-008-0274-z

Grossiord C (2020) Having the right neighbors: how tree species diversity modulates drought impacts on forests. New Phytol 228(1):42–49. https://doi.org/10.1111/nph.15667

Hagiwara T, Ishihara MI, Takabayashi J, Hiura T, Shiojiri K (2021) Effective distance of volatile cues for plant–plant communication in beech. Ecol Evol 11(18):12445–12452. https://doi.org/10.1002/ece3.7990

Halliday FW, Everingham SE, Bröcher M, Ebeling A, Kempel A, Mundim FM, Strauss AT, Xirocostas ZA, Kohli M (2025) Towards an integrative mechanistic framework for biodiversity–consumer relationships. Trends Ecol Evol 40(6):539–553. https://doi.org/10.1016/j.tree.2025.03.005

Hamilton J et al (2001) The carbon–nutrient balance hypothesis: its rise and fall. Ecol Lett 4(1):86–95. https://doi.org/10.1046/j.1461-0248.2001.00192.x

Heil M (2008) Indirect defence via tritrophic interactions. New Phytol 178(1):41–61. https://doi.org/10.1111/j.1469-8137.2007.02330.x

Herms DA, Mattson WJ (1992) The dilemma of plants: to grow or defend. Q Rev Biol 67(3):283–335. https://doi.org/10.1086/417659

Holopainen JK, Gershenzon J (2010) Multiple stress factors and the emission of plant VOCs. Trends Plant Sci 15(3):176–184. https://doi.org/10.1016/j.tplants.2010.01.006

Hou G, Delang CO, Lu X, Gao L (2020) A meta-analysis of changes in soil organic carbon stocks after afforestation with deciduous broadleaved, sempervirent broadleaved, and conifer tree species. Ann For Sci 77(4). https://doi.org/10.1007/s13595-020-00997-3

Huuskonen S, Domisch T, Finér L, Hantula J, Hynynen J, Matala J, Miina J, Neuvonen S, Nevalainen S, Niemistö P, Nikula A, Piri T, Siitonen J, Smolander A, Tonteri T, Uotila K, Viiri H (2021) What is the potential for replacing monocultures with mixed-species stands to enhance ecosystem services in boreal forests in Fennoscandia? For Ecol Manag 479:118558. https://doi.org/10.1016/j.foreco.2020.118558

Jactel H, Birgersson G, Andersson S, Schlyter F (2011) Non-host volatiles mediate associational resistance to the pine processionary moth. Oecologia 166(3):703–711

Jactel H, Moreira X, Castagneyrol B (2021) Tree diversity and forest resistance to insect pests: patterns, mechanisms and prospects. Annu Rev Entomol 66(1):277–296. https://doi.org/10.1146/annurev-ento-041720-075234

Jia S, Yang X, Castagneyrol B, Yang L, Yin Q, He C, Yang Z, Zhu Y, Hao Z (2023) Neighbouring tree effects on leaf herbivory: insect specialisation matters more than host plant leaf traits. J Ecol 112(1):189–199. https://doi.org/10.1111/1365-2745.14228

Jones IM, Koptur S (2015) Quantity over quality: light intensity, but not red/far-red ratio, affects extrafloral nectar production in Senna mexicana var. chapmanii. Ecol Evol 5(18):4108–4114. https://doi.org/10.1002/ece3.1644

Ju R-T, Gao L, Wei S-J, Li B (2017) Spring warming increases the abundance of an invasive specialist insect: links to phenology and life history. Sci Rep 7:14805. https://doi.org/10.1038/s41598-017-14989-3

Jucker T, Avăcăriței D, Bărnoaiea I, Duduman G, Bouriaud O, Coomes DA (2016) Climate modulates the effects of tree diversity on forest productivity. J Ecol 104(2):388–398. https://doi.org/10.1111/1365-2745.12522

Jucker T, Koricheva J, Finér L, Bouriaud O, Iacopetti G, Coomes DA (2020) Good things take time – diversity effects on tree growth shift from negative to positive during stand development in boreal forests. J Ecol 108(6):2198–2211. https://doi.org/10.1111/1365-2745.13464

Juntheikki M-R (1996) Comparison of tannin-binding proteins in saliva of Scandinavian and north American moose (*Alces alces*). Biochem Syst Ecol 24(7):595–601. https://doi.org/10.1016/S0305-1978(96)00084-1

Kant MR, Bleeker PM, Van Wijk M, Schuurink RC, Haring MA (2009) Plant volatiles in defence. In: VanLoon LC (ed) Plant innate immunity, vol 51. Academic Press Ltd-Elsevier Science Ltd., pp 613–666. https://doi.org/10.1016/S0065-2296(09)51014-2

Karban R, Shiojiri K (2009) Self-recognition affects plant communication and defense. Ecol Lett 12(6):502–506. https://doi.org/10.1111/j.1461-0248.2009.01313.x

Karban R, Huntzinger M, McCall AC (2004) The specificity of eavesdropping on sagebrush by other plants. Ecology 85(7):1846–1852. https://doi.org/10.1890/03-0593

Karban R, Wetzel WC, Shiojiri K, Pezzola E, Blande JD (2016) Geographic dialects in volatile communication between sagebrush individuals. Ecology 97(11):2917–2924. https://doi.org/10.1002/ecy.1573

Kessler A, Heil M (2011) The multiple faces of indirect defences and their agents of natural selection. Funct Ecol 25(2):348–357. https://doi.org/10.1111/j.1365-2435.2010.01818.x

Koricheva J (2002) Meta-analysis of sources of variation in fitness costs of plant antiherbivore defenses. Ecology 83(1):176–190

Koricheva J, Larsson S, Haukioja E, Keinänen M (1998) Regulation of woody plant secondary metabolism by resource availability: hypothesis testing by means of meta-analysis. Oikos 83(2):212–226. https://doi.org/10.2307/3546833

Koricheva J, Barton KE, Felix JA, Cooper A, Jensen J, Jucker T, Ruohomäki K (2025) Using long-term tree diversity experiments to explore the mechanisms of temporal shifts in forest ecosystem functioning. Oikos:e10872. https://doi.org/10.1111/oik.10872

Le Gall M, Behmer ST (2014) Effects of protein and carbohydrate on an insect herbivore: the vista from a fitness landscape. Integr Comp Biol 54(5):942–954. https://doi.org/10.1093/icb/icu102

Leong JV, Jorge LR, Seifert CL, Volf M (2022) Quantity and specialisation matter: effects of quantitative and qualitative variation in willow chemistry on resource preference in leaf-chewing insects. Insect Conserv Diver 15(4):453–460. https://doi.org/10.1111/icad.12559

Leong JV, Mezzomo P, Kozel P, Volfová T, de Lima Ferreira P, Seifert CL, Butterill PT, Freiberga I, Michálek J, Matos-Maraví P, Weinhold A, Engström MT, Salminen J-P, Segar ST, Sedio BE, Volf M (2024) Effects of individual traits vs. trait syndromes on assemblages of various herbivore guilds associated with central European Salix. Oecologia. https://doi.org/10.1007/s00442-024-05569-0

Liu Y, Singh SK, Pattanaik S, Wang H, Yuan L (2023) Light regulation of the biosynthesis of phenolics, terpenoids, and alkaloids in plants. Commun Biol 6(1):1–14. https://doi.org/10.1038/s42003-023-05435-4

Massey FP, Massey K, Press MC, Hartley SE (2006) Neighbourhood composition determines growth, architecture and herbivory in tropical rain forest tree seedlings. J Ecol 94(3):646–655. https://doi.org/10.1111/j.1365-2745.2006.01127.x

Mathesius U (2022) Humboldt review: are legumes different? Origins and consequences of evolving nitrogen fixing symbioses. J Plant Physiol 276:153765. https://doi.org/10.1016/j.jplph.2022.153765

Mattson WJ (1980) Herbivory in relation to plant nitrogen content. Annu Rev Ecol Syst 11(1):119–161. https://doi.org/10.1146/annurev.es.11.110180.001003

Mitchell RJ, Bellamy PE, Broome A, Ellis CJ, Hewison RL, Iason GR, Littlewood NA, Newey S, Pozsgai G, Ray D, Stockan JA, Stokes V, Taylor AFS (2022) Cumulative impact assessments of multiple host species loss from plant diseases show disproportionate reductions in associated biodiversity. J Ecol 110(1):221–231. https://doi.org/10.1111/1365-2745.13798

Moctezuma C, Hammerbacher A, Heil M, Gershenzon J, Méndez-Alonzo R, Oyama K (2014) Specific polyphenols and tannins are associated with defense against insect herbivores in the tropical oak *Quercus oleoides*. J Chem Ecol 40(5):458–467. https://doi.org/10.1007/s10886-014-0431-3

Monteith LG (1960) Influence of plants other than the food plants of their host on host-finding by tachinid parasites. Canadian Entomol 92(9):641–652. https://doi.org/10.4039/Ent92641-9

Moreira X, Abdala-Roberts L (2019) Specificity and context-dependency of plant–plant communication in response to insect herbivory. Curr Opin Insect Sci 32:15–21. https://doi.org/10.1016/j.cois.2018.09.003

Moreira X, Mooney KA, Zas R, Sampedro L (2012) Bottom-up effects of host-plant species diversity and top-down effects of ants interactively increase plant performance. Proc R Soc B Biol Sci 279(1746):4464–4472. https://doi.org/10.1098/rspb.2012.0893

Moreira X, Abdala-Roberts L, Parra-Tabla V, Mooney KA (2014) Positive effects of plant genotypic and species diversity on anti-herbivore defenses in a tropical tree species. PLoS One 9(8):e105438. https://doi.org/10.1371/journal.pone.0105438

Moreira X, Abdala-Roberts L, Rasmann S, Castagneyrol B, Mooney KA (2016) Plant diversity effects on insect herbivores and their natural enemies: current thinking, recent findings, and future directions. Curr Opin Insect Sci 14:1–7. https://doi.org/10.1016/j.cois.2015.10.003

Moreira X, Glauser G, Abdala-Roberts L (2017) Interactive effects of plant neighbourhood and ontogeny on insect herbivory and plant defensive traits. Sci Rep 7(1). https://doi.org/10.1038/s41598-017-04314-3

Moura RF, Del-Claro K (2023) Plants with extrafloral nectaries share indirect defenses and shape the local arboreal ant community. Oecologia 201(1):73–82. https://doi.org/10.1007/s00442-022-05286-6

Muiruri EW, Barantal S, Iason GR, Salminen J, Perez-Fernandez E, Koricheva J (2019) Forest diversity effects on insect herbivores: do leaf traits matter? New Phytol 221(4):2250–2260. https://doi.org/10.1111/nph.15558

Nguyen D, Castagneyrol B, Bruelheide H, Bussotti F, Guyot V, Jactel H, Jaroszewicz B, Valladares F, Stenlid J, Boberg J (2016) Fungal disease incidence along tree diversity gradients depends on latitude in European forests. Ecol Evol 6(8):2426–2438. https://doi.org/10.1002/ece3.2056

Niinemets Ü (2010) A review of light interception in plant stands from leaf to canopy in different plant functional types and in species with varying shade tolerance. Ecol Res 25(4):693–714. https://doi.org/10.1007/s11284-010-0712-4

Nybakken L, Lie MH, Julkunen-Tiitto R, Asplund J, Ohlson M (2018) Fertilization changes chemical defense in needles of mature Norway spruce (*Picea abies*). Front Plant Sci 9

Ormeño E, Fernandez C, Mévy J-P (2007) Plant coexistence alters terpene emission and content of Mediterranean species. Phytochemistry 68(6):840–852. https://doi.org/10.1016/j.phytochem.2006.11.033

Ortiz SK, Wolf AA (2024) Nitrogen-fixing plants increase soil nitrogen and neighbouring-plant biomass, but decrease community diversity: a meta-analysis reveals the mediating role of soil texture. J Ecol 112(10):2374–2385. https://doi.org/10.1111/1365-2745.14402

Pichersky E, Dudareva N (eds) (2020) Biology of plant volatiles, 2nd edn. CRC Press. https://doi.org/10.1201/9780429455612

Poeydebat C, Jactel H, Moreira X, Koricheva J, Barsoum N, Bauhus J, Eisenhauer N, Ferlian O, Francisco M, Gottschall F, Gravel D, Mason B, Muiruri E, Muys B, Nock C, Paquette A, Ponette Q, Scherer-Lorenzen M, Stokes V et al (2020) Climate affects neighbour-induced changes in leaf chemical defences and tree diversity–herbivory relationships. Funct Ecol 35(1):67–81. https://doi.org/10.1111/1365-2435.13700

Poorter H, Niinemets Ü, Poorter L, Wright IJ, Villar R (2009) Causes and consequences of variation in leaf mass per area (LMA): a meta-analysis. New Phytol 182(3):565–588. https://doi.org/10.1111/j.1469-8137.2009.02830.x

Poorter H, Niklas KJ, Reich PB, Oleksyn J, Poot P, Mommer L (2012) Biomass allocation to leaves, stems and roots: meta-analyses of interspecific variation and environmental control. New Phytol 193(1):30–50. https://doi.org/10.1111/j.1469-8137.2011.03952.x

Pretzsch H (2014) Canopy space filling and tree crown morphology in mixed-species stands compared with monocultures. For Ecol Manag 327:251–264. https://doi.org/10.1016/j.foreco.2014.04.027

Proß T, Bruelheide H, Potvin C, Sporbert M, Trogisch S, Haider S (2021) Drivers of within-tree leaf trait variation in a tropical planted forest varying in tree species richness. Basic Appl Ecol 50:203–216. https://doi.org/10.1016/j.baae.2020.11.001

Proß T, Bruelheide H, Haider S (2025) Within-individual leaf trait response to local light availability and biodiversity in a subtropical forest experiment. Ecology 106(7):e70160. https://doi.org/10.1002/ecy.70160

Radhika V, Kost C, Mithöfer A, Boland W (2010) Regulation of extrafloral nectar secretion by jasmonates in lima bean is light dependent. Proc Natl Acad Sci 107(40):17228–17233. https://doi.org/10.1073/pnas.1009007107

Reich PB (2014) The world-wide 'fast–slow' plant economics spectrum: a traits manifesto. J Ecol 102(2):275–301. https://doi.org/10.1111/1365-2745.12211

Reich PB, Walters MB, Ellsworth DS (1997) From tropics to tundra: global convergence in plant functioning. Proc Natl Acad Sci 94(25):13730–13734. https://doi.org/10.1073/pnas.94.25.13730

Riihimäki J, Kaitaniemi P, Koricheva J, Vehviläinen H (2005) Testing the enemies hypothesis in forest stands: the important role of tree species composition. Oecologia 142(1):90–97. https://doi.org/10.1007/s00442-004-1696-y

Roberts MR, Paul ND (2006) Seduced by the dark side: integrating molecular and ecological perspectives on the influence of light on plant defence against pests and pathogens. New Phytol 170(4):677–699. https://doi.org/10.1111/j.1469-8137.2006.01707.x

Rode M, Lemoine NP, Smith MD (2017) Prospective evidence for independent nitrogen and phosphorus limitation of grasshopper (*Chorthippus curtipennis*) growth in a tallgrass prairie. PLoS One 12(5):e0177754. https://doi.org/10.1371/journal.pone.0177754

Rosado-Sánchez S, Parra-Tabla V, Betancur-Ancona D, Moreira X, Abdala-Roberts L (2018a) Effects of tree species diversity on insect herbivory and leaf defences in *Cordia dodecandra*. Ecol Entomol 43(6):703–711. https://doi.org/10.1111/een.12648

Rosado-Sánchez S, Parra-Tabla V, Betancur-Ancona D, Moreira X, Abdala-Roberts L (2018b) Tree species diversity alters plant defense investment in an experimental forest plantation in southern Mexico. Biotropica 50(2):246–253. https://doi.org/10.1111/btp.12527

Roslin T, Salminen J-P (2008) Specialization pays off: contrasting effects of two types of tannins on oak specialist and generalist moth species. Oikos 117(10):1560–1568. https://doi.org/10.1111/j.0030-1299.2008.16725.x

Rozendaal DMA, Hurtado VH, Poorter L (2006) Plasticity in leaf traits of 38 tropical tree species in response to light; relationships with light demand and adult stature. Funct Ecol 20(2):207–216. https://doi.org/10.1111/j.1365-2435.2006.01105.x

Ruhnke H, Schädler M, Klotz S, Matthies D, Brandl R (2009) Variability in leaf traits, insect herbivory and herbivore performance within and among individuals of four broad-leaved tree species. Basic Appl Ecol 10(8):726–736. https://doi.org/10.1016/j.baae.2009.06.006

Rutten G, Hönig L, Schwaß R, Braun U, Saadani M, Schuldt A, Michalski SG, Bruelheide H (2021) More diverse tree communities promote foliar fungal pathogen diversity, but decrease infestation rates per tree species, in a subtropical biodiversity experiment. J Ecol 109(5):2068–2080. https://doi.org/10.1111/1365-2745.13620

Salminen J-P, Karonen M (2011) Chemical ecology of tannins and other phenolics: we need a change in approach. Funct Ecol 25(2):325–338. https://doi.org/10.1111/j.1365-2435.2010.01826.x

Salminen J-P, Roslin T, Karonen M, Sinkkonen J, Pihlaja K, Pulkkinen P (2004) Seasonal variation in the content of hydrolyzable tannins, flavonoid glycosides, and proanthocyanidins in oak leaves. J Chem Ecol 30(9):1693–1711. https://doi.org/10.1023/B:JOEC.0000042396.40756.b7

Sampaio T, Branco M, Guichoux E, Petit RJ, Pereira JS, Varela MC, Almeida MH (2016) Does the geography of cork oak origin influence budburst and leaf pest damage? For Ecol Manag 373:33–43. https://doi.org/10.1016/j.foreco.2016.04.019

Scartazza A, Di Baccio D, Bertolotto P, Gavrichkova O, Matteucci G (2016) Investigating the European beech (Fagus sylvatica L.) leaf characteristics along the vertical canopy profile: leaf structure, photosynthetic capacity, light energy dissipation and photoprotection mechanisms. Tree Physiol 36(9):1060–1076. https://doi.org/10.1093/treephys/tpw038

Schädler M, Jung G, Auge H, Brandl R (2003) Palatability, decomposition and insect herbivory: patterns in a successional old-field plant community. Oikos 103(1):121–132. https://doi.org/10.1034/j.1600-0706.2003.12659.x

Schnabel F, Beugnon R, Yang B, Richter R, Eisenhauer N, Huang Y, Liu X, Wirth C, Cesarz S, Fichtner A, Perles-Garcia MD, Hähn GJA, Härdtle W, Kunz M, Castro Izaguirre NC, Niklaus PA, von Oheimb G, Schmid B, Trogisch S et al (2025) Tree diversity increases forest temperature buffering via enhancing canopy density and structural diversity. Ecol Lett 28(3):e70096. https://doi.org/10.1111/ele.70096

Schuman MC (2023) Where, when, and why do plant volatiles mediate ecological signaling? The answer is blowing in the wind. Annu Rev Plant Biol 74:609–633. https://doi.org/10.1146/annurev-arplant-040121-114908

Shen P, Wang X, Zohner CM, Peñuelas J, Zhou Y, Tang Z, Xia J, Zheng H, Fu Y, Liang J, Sun W, Zhang Y, Wu C (2024) Biodiversity buffers the response of spring leaf unfolding to climate warming. Nat Clim Chang 14(8):863–868. https://doi.org/10.1038/s41558-024-02035-w

Siefert A, Violle C, Chalmandrier L, Albert CH, Taudiere A, Fajardo A, Aarssen LW, Baraloto C, Carlucci MB, Cianciaruso MV et al (2015) A global meta-analysis of the relative extent of intraspecific trait variation in plant communities. Ecol Lett 18(12):1406–1419. https://doi.org/10.1111/ele.12508

Smit C, Den Ouden J, Müller-Schärer H (2006) Unpalatable plants facilitate tree sapling survival in wooded pastures. J Appl Ecol 43(2):305–312. https://doi.org/10.1111/j.1365-2664.2006.01147.x

Staccone AP, Kou-Giesbrecht S, Taylor BN, Menge DNL (2021) Nitrogen-fixing trees have no net effect on forest growth in the coterminous United States. J Ecol 109(2):877–887. https://doi.org/10.1111/1365-2745.13513

Stamp N (2003) Out of the quagmire of plant defense hypotheses. Q Rev Biol 78(1):23–55. https://doi.org/10.1086/367580

Teshome DT, Zharare GE, Naidoo S (2020) The threat of the combined effect of biotic and abiotic stress factors in forestry under a changing climate. Front Plant Sci 11. https://doi.org/10.3389/fpls.2020.601009

Thompson PL, Isbell F, Loreau M, O'Connor MI, Gonzalez A (2018) The strength of the biodiversity–ecosystem function relationship depends on spatial scale. Proc R Soc B Biol Sci 285(1880):20180038. https://doi.org/10.1098/rspb.2018.0038

Tripathi S, Bhadouria R, Srivastava P, Devi RS, Chaturvedi R, Raghubanshi AS (2020) Effects of light availability on leaf attributes and seedling growth of four tree species in tropical dry forest. Ecol Process 9(1):2. https://doi.org/10.1186/s13717-019-0206-4

Trogisch S, Liu X, Rutten G, Xue K, Bauhus J, Brose U, Bu W, Cesarz S, Chesters D, Connolly J, Cui X, Eisenhauer N, Guo L, Haider S, Härdtle W, Kunz M, Liu L, Ma Z, Neumann S et al (2021) The significance of tree-tree interactions for forest ecosystem functioning. Basic Appl Ecol S1439179121000256. https://doi.org/10.1016/j.baae.2021.02.003

Underwood N, Inouye BD, Hambäck PA (2014) A conceptual framework for associational effects: when do neighbors matter and how would we know? Q Rev Biol 89(1):1–19. https://doi.org/10.1086/674991

Vanhakylä S, Salminen J-P (2023) Seasonal variation in plant polyphenols and related bioactivities across three years in ten tree species as visualized by mass spectrometric fingerprint mapping. Molecules 28(16):6093. https://doi.org/10.3390/molecules28166093

Vázquez-González C, Sampedro L, López-Goldar X, Solla A, Vivas M, Rozas V, Josefa Lombardero M, Zas R (2022) Inducibility of chemical defences by exogenous application of methyl jasmonate is long-lasting and conserved among populations in mature *Pinus pinaster* trees. For Ecol Manag 518:120280. https://doi.org/10.1016/j.foreco.2022.120280

Verdaguer D, Llorens L, Bernal M, Badosa J (2012) Photomorphogenic effects of UVB and UVA radiation on leaves of six Mediterranean sclerophyllous woody species subjected to two different watering regimes at the seedling stage. Environ Exp Bot 79:66–75. https://doi.org/10.1016/j.envexpbot.2012.01.008

Vivaldo G, Masi E, Taiti C, Caldarelli G, Mancuso S (2017) The network of plants volatile organic compounds. Sci Rep 7(1):11050. https://doi.org/10.1038/s41598-017-10975-x

Wang F, Deng J, Schal C, Lou Y, Zhou G, Ye B, Yin X, Xu Z, Shen L (2016) Non-host plant volatiles disrupt sex pheromone communication in a specialist herbivore. Sci Rep 6(1):32666. https://doi.org/10.1038/srep32666

Wang Z, Feng L, Frew A, Lu A, Yu Z, Huang Z (2024) Neighbourhood diversity effects on insect herbivory: plant leaf traits mediate associational resistance. J Ecol 112(11):2613–2623. https://doi.org/10.1111/1365-2745.14405

Westerband AC, Funk JL, Barton KE (2021) Intraspecific trait variation in plants: a renewed focus on its role in ecological processes. Ann Bot mcab011. https://doi.org/10.1093/aob/mcab011

Wetzel WC, Kharouba HM, Robinson M, Holyoak M, Karban R (2016) Variability in plant nutrients reduces insect herbivore performance. Nature 539(7629):425–427. https://doi.org/10.1038/nature20140

Williams LJ, Cavender-Bares J, Paquette A, Messier C, Reich PB (2020) Light mediates the relationship between community diversity and trait plasticity in functionally and phylogenetically diverse tree mixtures. J Ecol 108(4):1617–1634. https://doi.org/10.1111/1365-2745.13346

Wortemann R, Herbette S, Barigah T, Fumanal B, Alia R, Ducousso A, Gömöry D, Roeckel-Drevet P, Cochard H (2011) Genotypic variability and phenotypic plasticity of cavitation resistance in *Fagus sylvatica* L. across Europe. Tree Physiol 31:1175–1182. https://doi.org/10.1093/treephys/tpr101

Wright IJ, Reich PB, Westoby M, Ackerly DD, Baruch Z, Bongers F, Cavender-Bares J, Chapin T, Cornelissen JHC, Diemer M, Flexas J, Garnier E, Groom PK, Gulias J, Hikosaka K, Lamont BB, Lee T, Lee W, Lusk C et al (2004) The worldwide leaf economics spectrum. Nature 428:6985. https://doi.org/10.1038/nature02403

Wright AJ, Wardle DA, Callaway R, Gaxiola A (2017) The overlooked role of facilitation in biodiversity experiments. Trends Ecol Evol 32(5):383–390. https://doi.org/10.1016/j.tree.2017.02.011

Yue TX, Liu JY, Li ZQ, Chen SQ, Ma SN, Tian YZ, Ge F (2005) Considerable effects of diversity indices and spatial scales on conclusions relating to ecological diversity. Ecol Model 188(2):418–431. https://doi.org/10.1016/j.ecolmodel.2004.12.019

Zamora, R., Hódar, J., & Gomez, J. (1999). Plant–herbivore interaction: Beyond a binary vision. https://doi.org/10.1201/9781420007626.ch16

Zaynab M, Fatima M, Abbas S, Sharif Y, Umair M, Zafar MH, Bahadar K (2018) Role of secondary metabolites in plant defense against pathogens. Microb Pathog 124:198–202. https://doi.org/10.1016/j.micpath.2018.08.034

Zhang Q-H (2003) Interruption of aggregation pheromone in *Ips typographus* (L.) (Col. Scolytidae) by non-host bark volatiles. Agric For Entomol 5(2):145–153. https://doi.org/10.1046/j.1461-9563.2003.00173.x

Zhang Q-H, Schlyter F (2003) Redundancy, synergism, and active inhibitory range of non-host volatiles in reducing pheromone attraction in European spruce bark beetle Ips typographus. Oikos 101(2):299–310. https://doi.org/10.1034/j.1600-0706.2003.111595.x

Zhang Q-H, Schlyter F (2004) Olfactory recognition and behavioural avoidance of angiosperm nonhost volatiles by conifer-inhabiting bark beetles. Agric For Entomol 6(1):1–20. https://doi.org/10.1111/j.1461-9555.2004.00202.x

Zhang Q-H, Schlyter F (2010) Inhibition of predator attraction to kairomones by non-host plant volatiles for herbivores: a bypass-trophic signal. PLoS One 5(6):e11063. https://doi.org/10.1371/journal.pone.0011063

Zhu J, Sun L, He C, Cai Q, Zhu J, Ji C (2024) Large and thin leaves are compromised more by chewers. Ecosphere 15(1):e4748. https://doi.org/10.1002/ecs2.4748

Zu P, Boege K et al (2020) Information arms race explains plant-herbivore chemical communication in ecological communities. Science 368(6497):1377–1381. https://doi.org/10.1126/science.aba2965

Zvereva EL, Castagneyrol B, Kozlov MV (2024) Does spatial variation in insect herbivory match variations in plant quality? A meta-analysis. Ecol Lett 27(5):e14440. https://doi.org/10.1111/ele.14440

Plant Diversity Effects on Herbivory: Insights from Grassland Studies

3

Maximilian Bröcher, Sebastian T. Meyer, and Anne Ebeling

Abstract

Grasslands are globally important ecosystems that support high biodiversity and provide critical ecosystem services, yet they are increasingly threatened by land-use conversion and intensification. These pressures often reduce plant diversity, highlighting the need to understand the complex ecological interactions that sustain grassland functioning. Herbivory plays a central role by linking plants to higher trophic levels. Here, we review plant diversity effects on invertebrate herbivory in grasslands and point out commonalities and differences among studies, synthesizing these insights within the framework of existing theory, and highlight gaps in our understanding of those relationships. Although plant-herbivore interactions clearly depend on plant diversity, the direction and strength of these effects vary widely across the reviewed studies, reflecting their strong context dependency. Rather than a single dominant mechanism, multiple processes act simultaneously, with their relative importance influenced by the environmental conditions, the composition of plant and herbivore communities, trophic interactions, and management practices. To disentangle these multi-mechanistic and context-dependent relationships, future research must expand beyond Central European grasslands, since most grassland systems worldwide remain understudied. Progress also requires moving past a long-standing reliance on correlations and testing causality more directly. At the same time, incorporating the roles of other organisms that interact with both plants and herbivores will be essential for developing a holistic understanding of plant diversity-herbivory dynamics within the multitrophic networks that shape ecosystems.

M. Bröcher (✉) · A. Ebeling
Institute of Biodiversity, Ecology and Evolution, University of Jena, Jena, Germany
e-mail: maximilian.broecher@uni-jena.de

S. T. Meyer
Terrestrial Ecology Research Group, School of Life Sciences, Technical University of Munich, Freising, Germany

X. Moreira et al. (eds.), *Effects of Plant Diversity on Plant-Herbivore Interactions*, Ecological Studies 255, https://doi.org/10.1007/978-3-032-26461-9_3

3.1 Introduction

3.1.1 Significance of Grasslands

3.1.1.1 Definition and Distribution of Grasslands

Grasslands are characterized by the dominance of grasses and the absence or low abundance of woody vegetation (White et al. 2000; Gibson 2009). They can be broadly classified into natural grasslands, shaped by climate, fire, and wildlife grazing (e.g., prairies, savannas, steppe), and seminatural grasslands, which result from human management practices such as livestock grazing, hay-cutting, or burning. Globally, grasslands cover around 40% of Earth's terrestrial surface (White et al. 2000; Gibson 2009), exceeding the area of both forests and croplands. They are most extensive in sub-Saharan Africa, followed by Central Asia, with major distributions also in Europe, North America, and Oceania. Australia is especially notable, with more than 80% of its land area covered by grasslands, the largest grassland extent of any continent (White et al. 2000; Gibson 2009).

3.1.1.2 Ecological Importance

Grasslands play a vital role in the global carbon cycle and climate regulation, storing large amounts of soil carbon, comparable to—or even surpassing—that of forests (Gibson 2009). Depending on management practices, they can function as major carbon sinks, making them crucial for climate change mitigation (Lal 2004; Soussana et al. 2007; O'Mara 2012; Bengtsson et al. 2019). Grasslands also rank among the most biodiverse ecosystems on Earth, in some cases rivaling tropical rainforests in species richness. For instance, mountain grasslands in Argentina host up to 89 plant species per square meter, while semidry grasslands in Romania support 98 plant species within 10 square meters (Petermann and Buzhdygan 2021). In Europe, grasslands harbor about 18% of endemic vascular plants (Hobohm and Bruchmann 2009; Habel et al. 2013) and sustain more than two-thirds of the continent's butterfly species (WallisDeVries and Van Swaay 2009). Accordingly, they support a remarkably high diversity of insects and other invertebrates, including numerous specialists and regionally endemic species.

3.1.1.3 Economic and Social Importance

Grasslands are of high economic and social importance, supporting livestock industries through biomass production for meat, dairy, and wool, while also housing pollinators and natural predators that control agricultural pests (O'Mara 2012; Bengtsson et al. 2019). In Europe, they form part of long-standing cultural landscapes, attracting tourism and recreational activities like bird-watching and hiking. Globally, grasslands also carry deep spiritual and historical significance for indigenous peoples and rural communities (e.g., the North American prairies, South American Pampas, and Mongolian steppes). These communities draw on traditional knowledge, including sustainable grazing and controlled burning, to preserve the ecological health and resilience of grassland ecosystems (Bengtsson et al. 2019).

3.1.2 Invertebrate Herbivory in Grasslands

Compared to the estimated 4000 mammalian herbivore species worldwide (Ripple et al. 2015), invertebrate herbivores exceed them in both richness and biomass by multiple orders of magnitude (Grimaldi and Engel 2005; Bar-On et al. 2018). Roughly half of the more than 1 million insect species described worldwide are herbivorous (Grimaldi and Engel 2005), spanning major orders such as Lepidoptera (butterflies and moths), Coleoptera (beetles), Orthoptera (grasshoppers and crickets), and Hemiptera (true bugs). These invertebrates consume a wide range of plant parts using different feeding strategies, including leaf chewing, sap sucking, and mining (Price et al. 2011). Only a minority of herbivorous insects cause outbreaks or severe damage (despite a prominence of such examples in scientific literature); most persist at low densities. In grasslands, large-scale outbreaks are rare compared to forests, where outbreak herbivory can cause extensive defoliation. Instead, grassland herbivores exert chronic pressure through background herbivory—the persistent, low-intensity consumption of plant tissue that typically removes 1–3% of leaf area annually (Kozlov and Zvereva 2017). Despite these low rates, the collective impact of background herbivory is substantial across the plant community, and invertebrate herbivores substantially influence plant biomass, community composition, and nutrient cycling (Bigger and Marvier 1998; Kozlov and Zvereva 2017).

3.1.3 Ecological Implications of Background Herbivory

Background herbivory affects grassland ecosystems at multiple scales. At the individual level, herbivory reduces plant fitness by decreasing photosynthetic capacity through leaf area loss and reducing reproductive output through seed predation (Agrawal 1999; McArt et al. 2013). At the community level, herbivory can decrease overall biomass or shift species composition by suppressing dominant plants and enabling less competitive species to persist (Bröcher et al. 2023). By preventing competitive exclusion, these dynamics may enhance community resilience to environmental fluctuations. Herbivory also affects plant growth indirectly by accelerating nutrient cycling through decomposition of plant matter (Nitschke et al. 2015) and by reducing light limitation through canopy thinning and litter removal (Borer et al. 2014; Eskelinen et al. 2022).

3.1.4 Drivers of Invertebrate Herbivory

Multiple factors influence herbivory rates in grasslands (Kozlov and Zvereva 2017; Halliday et al. 2025). Abiotic conditions such as temperature and precipitation affect both herbivore activity and plant defenses (Hamann et al. 2021; Lin et al. 2023). Management practices including mowing and grazing alter habitat structure and resource availability for herbivores (Gossner et al. 2014; Shinohara et al. 2019; Künast et al. 2025). Plant community characteristics, particularly species composition and diversity, shape herbivore communities through variation in plant palatability, nutritional quality, and chemical defenses (Loranger et al. 2012; Ebeling et al. 2018b). Understanding how plant diversity affects herbivory is critical for predicting ecosystem functioning in grasslands and for informing conservation and restoration strategies.

3.1.5 Scope of This Review

As global biodiversity faces unprecedented shifts, understanding the role of altered diversity within terrestrial plant communities is becoming increasingly important. Variations in plant diversity have profound implications for ecological interactions, particularly in grassland ecosystems where changes in plant composition can strongly influence invertebrate herbivory.

We conducted a systematic review of empirical studies addressing this relationship to better understand how shifts in grassland plant diversity affect ecosystem processes. To identify relevant studies, we used the following Boolean search query in the Web of Science database:

> *grassland AND ("plant richness" OR "biodiversity experiment" OR "plant diversity" OR "plant species richness" OR "number of plant species") AND (herbiv* OR "leaf damage" OR "plant damage")*

This search yielded 721 manuscripts (as of February 2025), which were screened for inclusion. We retained only studies that (i) investigated invertebrates as agents of herbivory, and (ii) measured herbivory directly or via exclusion of herbivores (e.g., spraying, fencing, mesocosms, etc.). We excluded studies that examined only herbivore abundance in relation to plant richness, because abundance can be a very poor predictor of herbivory rates (Ebeling et al. 2014; Croijmans et al. 2025). We also excluded studies using herbivory as an explanatory variable because of our focus on plant species richness effects. After screening, 43 studies remained, encompassing both experimental and observational approaches, species- and community-level measurements, and a broad range of herbivory metrics (Fig. 3.1, Table 3.1). Experimental studies actively manipulated plant species richness, while observational studies relied on existing (natural) gradients. Whereas species-level herbivory focuses on damage to a single plant species, community-level herbivory captures damage across the plant community. Community-level herbivory is often derived as the sum of species-level herbivory weighting by the relative abundance of each species in the community.

In the following sections, we synthesize insights from these empirical studies alongside theoretical frameworks to provide a comprehensive overview of how plant diversity shapes invertebrate herbivory in grasslands. We highlight both commonalities and differences among studies, discuss their broader implications, and identify open questions to guide future research.

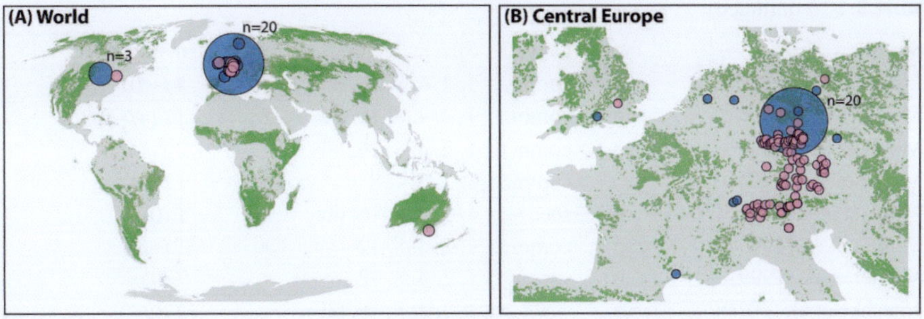

Fig. 3.1 Global distribution of studies investigating the effects of plant diversity on herbivory in grasslands worldwide (**a**) and with a detailed view of Central Europe (**b**). Grassland extent is shown in green (Parente et al. 2024). Locations are indicated by points, colored by study type (observational = pink; experimental = blue). Observational studies often include multiple sampling sites, resulting in several points per study (Fischer et al. 2012; Hülber et al. 2015; Fricke et al. 2022), whereas experimental studies are usually restricted to a single site. Point size reflects the number of studies conducted at the same location (range: 1–20)

Table 3.1 Summary of published studies included in the literature review, examining plant diversity effects on herbivory in experimental (exp.) or observational (obs.) settings, at either the species (spec.) or plant communities (comm.) level

Study	Type	Level	Species richness gradient	n. effects
Barnes et al. (2020)	exp.	comm.	1, 2, 4, 8, 16	2 (+)
Bröcher et al. (2024)	exp.	comm.	1, 2, 4, 8, 16, 60	12 (+)
Bröcher et al. (2023)	exp.	spec.	1, 2, 3, 4, 8	24 (+/−)
Cappelli et al. (2022)	exp.	spec.	1, 4, 8, 20	18 (+/−)
Dinnage (2013)	obs.	comm.	3 to 18	1 (+)
Ebeling et al. (2014)	exp.	comm.	1, 2, 4, 8, 16, 60	1 (+)
Egorov et al. (2017)	obs.	comm.	13.3 (mean) to 55.6 (mean)	3 (+/0/−)
Fischer et al. (2012)	obs.	comm.	18 to 80	1 (0)
Fricke et al. (2022)	obs.	comm.	12 to 52	3 (+/0/−)
Guo et al. (2016)	exp.	spec.	1 vs. 4	1 (0)
Gutiérrez et al. (2020)	exp.	comm.	2 vs. 8	1 (+)
Halliday et al. (2023)	obs.	comm.	7 to 30	1 (0/−)
Hanley (2004)	exp.	comm.	3, 6, 12	1 (+/0)
Hanley (2004)	exp.	spec.	3, 6, 12	3 (0)
Hudewenz et al. (2012)	exp.	comm.	1, 2, 4, 8, 16, 60	1 (0)
Hülber et al. (2015)	obs.	spec.	27 to 57	1 (+)
Kigathi et al. (2019)	exp.	spec.	1, 2, 3	1 (+/0)
Korell et al. (2016)	exp.	comm.	1, 6	1 (0)
Lanta (2007)	exp.	comm.	1, 2, 4, 6	1 (0/−)
Lau et al. (2008)[a]	exp.	spec.	1, 16	6 (+/0/−)
Li et al. (2024)	exp.	comm.	1, 2, 4, 8, 16, 60	1 (+)
Lipowsky et al. (2011)	exp.	spec.	1, 60	5 (+/0/−)

(continued)

Table 3.1 (continued)

Study	Type	Level	Species richness gradient	n. effects
Loranger et al. (2014)	exp.	comm.	1, 2, 4, 8, 16, 60	4 (+/0)
Meyer et al. (2016)	exp.	comm.	1, 2, 4, 8, 16, 60	17 (+/−)
Meyer et al. (2017)	exp.	comm.	1, 2, 4, 8, 16, 60	20 (+)
Meyer et al. (2017)	exp.	comm.	1, 2, 3, 4, 8	12 (+)
Mraja et al. (2011)	exp.	spec.	1, 2, 4, 8, 16, 60	1 (0)
Mulder et al. (1999)	exp.	comm.	1, 2, 4, 8, 12	1 (+/0)
Mulder et al. (1999)	exp.	spec.	1, 2, 4, 8, 12	7 (+)
Nitschke et al. (2010)	exp.	spec.	1, 2, 4, 8, 16, 60	2 (0/−)
Nitschke et al. (2015)	exp.	spec.	1, 2, 3, 6,	1 (0)
Nitschke et al. (2017)	exp.	spec.	1, 2, 4, 8	3 (0/−)
Petermann et al. (2010)	exp.	comm.	1, 2, 3, 4, 6, 9	1 (unimodal)
Pfisterer et al. (2003)	exp.	comm.	1, 2, 4, 8, 32	1 (0/−)
Prieur-Richard et al. (2002)	exp.	spec.	3, 6, 18	2 (+)
Reidinger et al. (2012)	exp.	spec.	0, 4, 15	4 (+/0/−/unimodal)
Ristok et al. (2023)[a]	exp.	comm.	1, 2, 4, 8	6 (+/0/−)
Rossetti et al. (2019)	obs.	spec.	2 to 21	2 (+/−)
Scherber et al. (2006b)	exp.	comm.	1, 2, 4, 8, 16, 60	4 (+/0)
Scherber et al. (2006b)	exp.	spec.	1, 2, 4, 8, 16, 60	4 (+/0/−)
Scherber et al. (2010)	exp.	comm.	1, 2, 4, 8, 16, 60	2 (0)
Scherber et al. (2006a)	exp.	spec.	1, 2, 4, 8, 16, 60	2 (0/−)
Seabloom et al. (2017)	exp.	comm.	1, 4, 16	1 (+/0)
Silvertown et al. (2006)	obs.	spec.	10 to 37	5 (−)
Stein et al. (2010)	obs.	comm.	13 to 38	1 (−)
Unsicker et al. (2006)	obs.	comm.	18 to 45	3 (0/−)
Unsicker et al. (2006)	obs.	spec.	18 to 45	4 (+/0)
Wilson (1994)	obs.	spec.	4.1 (mean), 5.1 (mean)	2 (+/0)

The table reports the gradient of plant species richness over which the relationship was measured, the total number of tested plant diversity-herbivory effects and their directions, i.e., positive (+), neutral (0), negative (−), and unimodal relationships. The total number of positive, neutral, and negative effects is summarized in Fig. 3.2

[a]Study was not detected by the search query and was added manually

3.2 Effects of Grassland Plant Diversity on Herbivory

We aimed to identify general patterns in plant diversity effects on herbivory in grasslands. However, our systematic literature review revealed high heterogeneity in results, regardless of whether studies were experimental or observational and whether they examined species-level or community-level effects. Given this heterogeneity, we first provide an overview of documented relationships and explanations for the observed differences. We then discuss the underlying mechanisms for biodiversity effects.

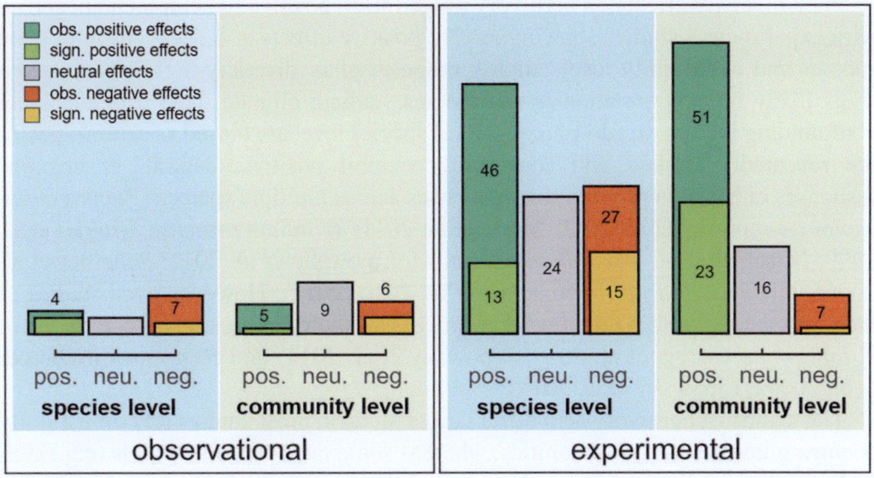

Fig. 3.2 The number of plant diversity–herbivory relationships grouped by study setting (observational vs. experimental) and by the level of analysis (community vs. species level). Each relationship represents a plant diversity treatment effect on a measure of herbivory. When studies include multiple measures of herbivory, multiple time points, several experimental diversity gradients, or multiple plant species, each is counted separately, as results may differ among them. The direction of the plant diversity effect is indicated by color: green for positive, gray for neutral, and orange for negative relationships. Plant diversity effects are shown either as observed (obs.) or as statistically tested (sign)

The higher number of effect sizes reported in experimental compared to observational studies likely reflects differences in study design and methodology (Fig. 3.2). Although observational studies often span multiple sites (Egorov et al. 2017), they are rarely repeated across years or seasons, unlike many experimental studies, which frequently include temporal replication and thus contribute multiple, nonindependent effects. Plant diversity effects are generally consistent across seasons (Scherber et al. 2010; Loranger et al. 2014; Meyer et al. 2017), and although they can vary considerably between years, the direction of these effects rarely changes (Mulder et al. 1999; Stein et al. 2010; Scherber et al. 2010; Bröcher et al. 2024). Furthermore, observational studies often rely on incidence-based measures (e.g., the presence or absence of damage), which are suitable for natural field conditions but provide lower precision (Wilson 1994; Hülber et al. 2015). Experimental studies, on the other hand, typically use more precise metrics such as proportional leaf damage (damaged or consumed area relative to total leaf area; Scherber et al. 2006a; Lipowsky et al. 2011; Guo et al. 2016), biomass consumption estimates (Mulder et al. 1999; Bröcher et al. 2023), or energy flux-based measures (Barnes et al. 2020; Ristok et al. 2023), resulting in a higher number of quantifiable effect sizes. Nonetheless, findings tend to be consistent within each herbivory metric (Loranger et al. 2014; Ebeling et al. 2014; Meyer et al. 2017; Bröcher et al. 2024).

Plant diversity effects on herbivory ranged from positive to neutral and negative, with experimental studies showing mostly positive effects at the level of individual species and community level. Such a range in plant diversity-herbivory relationships likely reflects variation in the species studied, climate, land use, and other confounding factors. In addition, results at species level are biased as certain species are repeatedly studied and consistently exhibit positive, neutral, or negative responses rather than representing responses across multiple species (*Plantago lanceolata*, Lipowsky et al. 2011; Mraja et al. 2011; *Trifolium pratense*, Mulder et al. 1999; Nitschke et al. 2015; *Rumex acetosa*, Lipowsky et al. 2011; Scherber et al. 2006a, b; *Centaurea jacea*, Nitschke et al. 2010, 2017). However, even studies of the same species within the same experiment can show contrasting outcomes, as for *P. lanceolata* (Mraja et al. 2011; Lipowsky et al. 2011) and *R. acetosa* (Scherber et al. 2006a; Lipowsky et al. 2011).

The agents of herbivory also differ across studies; most studies rely on naturally occurring invertebrate communities, whereas some target specific guilds (e.g., generalist and specialist herbivores; Lau et al. 2008) or taxa such as mollusks (Hanley 2004; Korell et al. 2016), grasshoppers (Pfisterer et al. 2003; Lanta 2007; Nitschke et al. 2015), tephritid flies (Nitschke et al. 2017), aphids (Petermann et al. 2010), or caterpillars (*Spodoptera littoralis*; Kigathi et al. 2019). Interestingly, negative effects observed in experimental studies at the levels of communities tend to occur only when specific herbivore groups, rather than the entire herbivore community, are considered (e.g., grasshoppers: Pfisterer et al. 2003; Lanta 2007; aphids: Petermann et al. 2010; mollusks: Korell et al. 2016).

Observational studies, when conducted across broad spatial scales, may confound plant diversity effects with environmental gradients such as community composition (Wilson 1994; Unsicker et al. 2006; Fischer et al. 2012; Egorov et al. 2017; Fricke et al. 2022; Halliday et al. 2025), elevation (Hülber et al. 2015), plant productivity (Unsicker et al. 2006; Stein et al. 2010; Fischer et al. 2012), land use (Fischer et al. 2012; Hülber et al. 2015; Egorov et al. 2017; Rossetti et al. 2019), elevation and temperature (Halliday et al. 2023), as well as plant phylogenetic diversity (Dinnage 2013; Egorov et al. 2017; Halliday et al. 2023). Combined with the currently small number of reported relationships, this limits our ability to draw general conclusions from the literature for observational studies.

Even though studies vary widely in design and focus, certain patterns emerged from our review. The strong influence of the Jena Experiment represents both a caveat and an opportunity: 19 of the 43 studies included were conducted at this single site, contributing 124 of the 200 observed relationships. This concentration may bias our understanding of plant diversity effects on herbivory in grasslands, as patterns observed at the Jena Experiment may not generalize to other grassland systems. However, it also provides a unique advantage by holding constant confounding factors such as soil conditions, plant communities, and herbivore assemblages, allowing a clearer assessment of plant diversity effects.

3.3 Underlying Mechanisms

The range of patterns described above raises the question: What mechanisms drive these effects of plant diversity on herbivory? At least a dozen hypotheses have been proposed (summarized in Halliday et al. 2025), all based on the principle that changes in plant community diversity affect both individual plants and the broader plant community. Key factors include total plant biomass (section *Resource availability*), focal plant density (section *Plant density effects*), plant traits (section *The role of plant traits*), and neighborhood effects and resource diversity (section *Plant community composition*) (Halliday et al. 2025). These changes, in turn, influence herbivores (section *Herbivores*) and their natural enemies (section *Predators*), affecting their abundance, species richness, and performance, and ultimately the extent of plant damage (Ebeling et al. 2018a). Within each section, we first describe ecological theory on the corresponding mechanism and then summarize empirical evidence from grassland plant diversity studies (results from the literature search are highlighted in bold and listed in Table 3.1).

3.3.1 Resource Availability

At the level of the whole plant community, higher plant species richness increases plant productivity (Cardinale et al. 2011), thus, providing more plant resources that can support higher herbivore abundances. According to the *More Individual Hypothesis* (Srivastava and Lawton 1998), a larger herbivore community can sustain more species that exploit available resources more completely, potentially leading to increased herbivory.

Repeatedly, grassland studies have demonstrated higher abundance and diversity of herbivores with increasing plant diversity (e.g., Haddad et al. 2001; Ebeling et al. 2018a). Consistent with the predictions of the more individual hypothesis, several studies from the reviewed literature report that herbivores consume more leaf material at higher plant diversity (Loranger et al. 2014; Meyer et al. 2017; Bröcher et al. 2023, 2024), although these increases do not always exceed increases in plant productivity. For instance, Loranger et al. (2014) found that higher plant biomass at higher plant species richness resulted in more leaf biomass consumed, but the percentage of herbivory damage increased only marginally.

3.3.2 Plant Density Effects

Productivity often drives herbivory at the community level, but plant density is especially relevant at the species level. The *Resource Concentration* (Root 1973) and *Resource Dilution* (Otway et al. 2005) hypotheses describe how specialist herbivores respond to host plant availability. The *Resource Concentration Hypothesis* predicts higher herbivory where host plants are dense, such as in monocultures, because herbivores can more easily locate and remain on their hosts. In contrast, the *Resource Dilution Hypothesis* predicts higher per-plant herbivory in more diverse (and thus host-sparse) communities, where herbivores are forced to concentrate on fewer hosts. These represent positive and negative density-dependent herbivore responses, respectively, and primarily describe herbivore movement and distribution.

Empirical evidence for these opposing density-dependent responses is mixed (Unsicker et al. 2006; Hülber et al. 2015). Even within a single study, Bröcher et al. (2023) found that species-specific herbivory varied from strongly negative to positive with host biomass, supporting both hypotheses depending on species identity and trait composition (see section *The role of plant traits*). Moreover, density effects often interact with associational effects, mediated by the properties of neighboring plants that influence herbivore damage to focal individuals (see section *Plant community composition*).

3.3.3 The Role of Plant Traits

Plant traits explain variability in herbivory by determining plant palatability through their nutritional and defensive properties (Hanley et al. 2007; Carmona et al. 2011; Loranger et al. 2012, 2013; Těšitel et al. 2021). Architectural traits such as height, leaf area, and growth form influence microclimate, shelter, and resource availability to herbivores (Roscher et al. 2012; Coverdale and Davies 2023). Temporal traits, including phenology, regrowth capacity, and longevity, determine the continuity and predictability of food resources. Both individual plant traits and community-level mean traits and trait diversity can shift when plants grow with heterospecifics, causing variation in plant diversity–herbivory relationships.

Traits linked to plant quality are among the most consistent and tested predictors of herbivory. Higher nitrogen concentration (Mraja et al. 2011; Gutiérrez et al. 2020; Bröcher et al. 2023), higher specific leaf area (Mraja et al. 2011), lower C:N ratio (Ebeling et al. 2014), and lower leaf dry matter content (Bröcher et al. 2023) generally increased herbivory at higher plant diversity. However, their effects vary among systems, and single traits can have opposing outcomes; for instance, high nitrogen in *Plantago lanceolata* was associated with both increased herbivory and stronger chemical defenses (Mraja et al. 2011). Plant architecture and traits related to resource availability received limited support as predictors of herbivory. For example, the number of leaves showed only weak relationships with herbivory (Lipowsky et al. 2011), while community leaf area index (LAI) produced mixed results, i.e., plant diversity enhanced both LAI and herbivory in *Plantago lanceolata* but not in *Rumex acetosa* (Scherber et al. 2006a, 2006b). Plant height generally increased herbivory but did not consistently mediate diversity effects (Lipowsky et al. 2011; Loranger et al. 2014; Rossetti et al. 2019). Increases in these traits with plant diversity may enhance resource availability and thereby herbivory, yet they are often not measured directly but instead captured indirectly through biomass production (Bröcher et al. 2023; see section *Resource availability*). Traits such as phenology, regrowth capacity, and longevity influence the temporal continuity of food resources but are rarely tested as mediators of plant diversity effects on herbivory (but see Mraja et al. 2011).

3.3.4 Plant Community Composition

Plant community composition shapes the effects of plant diversity on herbivory through both increased diversity of plant traits and synergistic interactions among them. Greater differences among plants increase resource heterogeneity, allowing herbivores to exploit a broader range of resources and maintain more diverse populations (*Resource Heterogeneity Hypothesis*; Hutchinson 1959; MacArthur 1972), while the specific combination of species and the ways their traits interact within the community further modulate these effects. For instance, higher leaf nitrogen content may enhance herbivory, but its impact depends on the relative nitrogen levels of neighboring plants. Such variation in traits among neighboring species underlies neighborhood contrast (Alm Bergvall et al. 2006) and associational effects (Barbosa et al. 2009), where plants with high nutritional quality or low defenses can attract herbivores either to themselves (neighborhood contrast susceptibility) or to the entire community (associational susceptibility), while the opposite patterns generate neighborhood contrast defense and associational resistance.

Evidence from the reviewed studies supports the idea that plant community composition influences plant diversity–herbivory relationships (e.g., Prieur-Richard et al. 2002; Unsicker et al. 2006; Dinnage 2013; Loranger et al. 2014; Fricke et al. 2022). In several cases, phylogenetic diversity even outperforms species richness as a predictor of herbivory (Egorov et al. 2017; Halliday et al. 2023), likely because grouping plants by traits or evolutionary relatedness (i.e., functional or phylogenetic diversity) captures more biologically meaningful variation. However, the mechanistic understanding behind these patterns remains limited, as they often depend on variables that are conceptually simple but empirically difficult to quantify. For instance, resource diversity is likely a key driver of plant diversity effects on herbivory mediated by community composition, yet it is nearly impossible to measure, because it depends on which resources are relevant to particular herbivores. For similar reasons, compositional effects beyond the number or presence of functional groups have rarely been explored. To date, only one study has shown that less palatable species benefit from more palatable neighbors through contrast defense (Bröcher et al. 2023), although similarly indirect inferences can also be drawn from studies examining the presence–absence effects of functional groups.

3.3.5 Herbivores

Plant diversity influences herbivore communities (e.g., richness, abundance, trait composition) by altering plant traits, resource availability, resource diversity, structural complexity (see sections above), or top-down control by predators (section *Predators*). At a finer scale, individual herbivores could adjust their behavior in response to variation in plant diversity with consequences for individual performance. Diverse plant communities enable herbivores to dilute toxins and better meet their stoichiometric demands (Bernays and Bright 1993), while in monocultures, herbivores often compensate for unbalanced diets by consuming more plant material (compensatory feeding: Yang and Joern 1994). Despite their central role in plant diversity-herbivory relationships, herbivores are frequently overlooked, leaving unresolved whether increased herbivory results from higher herbivore abundance, shifts in community composition, or altered behavior.

Many of the reviewed studies demonstrated that greater plant diversity is associated with higher herbivore abundance and richness, as well as increased herbivory damage (Ebeling et al. 2014; Barnes et al. 2020; Bröcher et al. 2024). Sometimes nonlinear relationships have been reported, such as for aphid abundance, where opposing effects produce peaks at intermediate plant diversity (Petermann et al. 2010; Reidinger et al. 2012). However, plant diversity effects on herbivory depend on feeding specialization and feeding types of the herbivores. Specifically, specialized herbivores, chewing, sucking and mining species consumed less, and generalist herbivores consumed more in diverse plant communities compared to monocultures (Lau et al. 2008; Rossetti et al. 2019; Halliday et al. 2023). Most of these findings rely on indirect indicators of herbivores, specifically the presence and characteristics of feeding marks, not on direct estimates of herbivore abundance (Fischer et al. 2012; Rossetti et al. 2019; Halliday et al. 2023).

The role of species behavior in driving the effects of plant diversity on herbivory is rarely studied and is in most cases only assessed indirectly by measuring the performance of single herbivore species. Overall, there is no evidence that plant diversity-mediated changes in species performance translate into changes in overall herbivory (Pfisterer et al. 2003; Nitschke et al. 2015; Gutiérrez et al. 2020).

3.3.6 Predators

Beyond bottom-up effects, herbivore abundance and behavior are also influenced by predators. Predators not only reduce herbivore numbers through direct consumption but can also alter their behavior by increasing perceived predation risk. According to the *Enemies Hypothesis*, predators benefit from the greater structural complexity of diverse plant communities more than from the higher plant species richness itself (Russell 1989). The higher complexity provides more niches, foraging opportunities, and enemy-free space (shelter) for predators, which in turn might increase the suppression of herbivore populations.

Many reviewed studies report positive relationships between plant diversity and both herbivores and their natural enemies (e.g., Unsicker et al. 2006; Nitschke et al. 2017; Bröcher et al. 2024; but see Petermann et al. 2010). Additional support for the *Enemies Hypothesis* comes from Barnes et al. (2020), who reported stronger top-down control in more diverse plant communities, where increased plant diversity reduced herbivore impacts on plants by simultaneously benefitting predators. However, as it is difficult to quantify top-down control without, e.g., directly manipulating predators, direct evidence that top-down control strengthens with increasing plant diversity remains limited in grassland studies.

3.4 Future Directions

Despite extensive research on plant diversity effects on herbivory in grasslands, a general answer to the question of whether herbivory increases, decreases, or remains unaffected by plant diversity remains elusive. Reported effects range from positive to neutral to negative, reflecting strong context dependency. Nevertheless, our review revealed a slight overall tendency toward positive diversity effects on herbivory. Local environmental conditions, plant community composition, herbivore communities, trophic interactions, and management practices all interact to determine the direction and magnitude of diversity effects. To advance understanding and develop generalizable insights into when and how plant diversity influences herbivory, several critical research gaps must be addressed. In the following section, we will outline potential directions for future research to address these questions.

3.4.1 Complexity of Mechanisms

Similar to the mechanisms for plant diversity effects in general (Eisenhauer et al. 2024), the evidence reviewed suggests that plant diversity effects on herbivory rates are driven by multiple mechanisms acting in concert rather than a single dominant factor. Most research to date has focused on subsets of potential drivers (resource availability, plant traits, community composition, herbivores, or predators) while neglecting other components of the causal chain (Ebeling et al. 2014; Gutiérrez et al. 2020; Ristok et al. 2023; Bröcher et al. 2023). Although the simultaneous action of multiple mechanisms is widely recognized (Ebeling et al. 2014; Halliday et al. 2023, 2025; Bröcher et al. 2023), and although a conceptual framework for testing multiple mechanisms was recently developed (Halliday et al. 2025), it remains untested. Furthermore, most studies infer mechanisms from correlations rather than directly testing causality (e.g., Scherber et al. 2006b; Rossetti et al. 2019; Bröcher et al. 2023). Future studies should advance the field by incorporating multiple mechanisms simultaneously to better understand these complex ecological interactions and shift the focus from correlation toward causal inference to better understand observed herbivory responses and their underlying mechanisms.

3.4.2 Context Dependency

The review of existing studies shows that plant diversity effects on herbivory are highly context dependent. Both the magnitude and direction of these effects, as well as the mechanisms underlying them, vary across ecological contexts. Key factors include plant community characteristics, such as the dominant plant species and their functional traits (Prieur-Richard et al. 2002; Mraja et al. 2011; Bröcher et al. 2023); herbivore identity, including dominant groups and feeding modes (Lau et al. 2008; Halliday et al. 2023); trophic interactions, such as the presence of predators or pathogens (Fischer et al. 2012; Barnes et al. 2020); and environmental and management conditions, including grassland type (Wilson 1994; Fischer et al. 2012), land-use intensity (Egorov et al. 2017), management practices (Hudewenz et al. 2012), and climate (Halliday et al. 2023). Most studies have not examined these contexts simultaneously, yet this approach is necessary to determine their relative importance in shaping the plant diversity-herbivory relationship. Many of these factors are not easily separable, for example, land use intensity and plant community composition (Egorov et al. 2017), or grassland type and climatic conditions (Fischer et al. 2012; Halliday et al. 2023). To identify the most important global drivers of the plant diversity-herbivory relationship, standardized studies are needed that test gradients in plant community composition and functional diversity, herbivore community structure, trophic interactions, and environmental conditions, including land use and climate.

3.4.3 Network Complexity

Studies on plant diversity-herbivory relationships in grasslands often overlook interactions among herbivores and between other trophic groups, such as pathogens, predators, and decomposers. These interactions may be crucial for explaining variation in herbivory across plant diversity gradients. Promising, but poorly studied, are interactions among herbivore groups (e.g., generalists vs. specialists: Lau et al. 2008) and interactions between above- and belowground herbivores (Stein et al. 2010). Interactions between invertebrates and vertebrate grazers, or between herbivores and pathogens, also remain largely unexplored in grassland plant diversity-herbivory research (but see Fischer et al. 2012), even though evidence suggests that these groups can exert complementary or competitive effects (Kluth et al. 2002; Gómez and González-Megías 2007; Huntzinger et al. 2008). For predators, theory suggests that their interactions with herbivores can shape plant diversity-herbivory relationships (section *Predators*; Russell 1989; Snyder et al. 2006; Terborgh 2015), but existing studies have largely relied solely on correlative approaches (Unsicker et al. 2006; Nitschke et al. 2017). Future research should move beyond a purely bottom-up perspective toward a holistic approach that identifies which groups drive herbivory in grasslands, their natural enemies, and the roles of other trophic interactions (e.g., pathogens, decomposers) while accounting for the dynamics of these interactions, which can strongly influence plant diversity-herbivory relationships.

3.4.4 Herbivory Effects on Temporal Dynamics

Herbivory shapes grassland ecosystems across multiple temporal scales. While herbivory is often associated with reduced plant fitness, plant responses depend on strategies such as avoidance, tolerance, or defense (Gong and Zhang 2014). For example, palatable forbs may remain abundant not despite, but because of, their strategy to tolerate damage, making herbivory a cost-efficient trade-off for maximizing reproduction (Bröcher et al. 2023). Considering plant strategies helps clarify how herbivory shapes community composition, including successional dynamics and forage quality, both influenced by selective feeding. These compositional shifts, in turn, alter the context of herbivory, feeding back into the mechanisms that drive plant diversity-herbivory relationships (Schmitz et al. 2000; Bröcher et al. 2025). Such changes can cascade to affect decomposition and nutrient cycling, ultimately shaping plant growth and long-term community dynamics (Ayal 2007; Holt and Barfield 2013). Reciprocal relationships like these are central to temporal patterns of plant diversity and herbivory in grasslands but remain understudied (Bröcher 2025). This topic represents a promising avenue for future research, as these dynamics are a fundamental, yet often overlooked, driver of ecosystem function. Even coevolutionary processes can be understood in this context: Interspecific variation drives shifts in community composition, while intraspecific variation enables evolutionary responses to selective herbivory.

3.5 Conclusions

Effects of plant species richness on arthropod herbivory in grasslands in the published literature are variable ranging from positive over neutral to positive with a slight overall tendency toward positive diversity effects on herbivory. Multiple mechanisms act simultaneously to determine plant diversity effects on herbivory, but their relative contributions depend on local context. Grassland management should be adaptive and tailored to these local ecological contexts, recognizing that plant diversity does not uniformly regulate herbivory but interacts with environmental conditions and functionally diverse consumer communities. Currently, research is dominated by studies from Central Europe, leaving large portions of global grasslands understudied (Fig. 3.1). Even in well-studied grasslands, we are only beginning to partition the net effect of plant diversity on herbivory into components that can be compared across contexts. A holistic perspective is needed, considering not only herbivores but also other organismal groups that both affect and are affected by herbivory, which can, in turn, influence plant-herbivore interactions. These feedbacks are essential to understanding ecological dynamics in grasslands and, more broadly, one of the most central organismal interactions on earth—herbivory.

References

Agrawal AA (1999) Induced responses to herbivory in wild radish: effects on several herbivores and plant fitness. Ecology 80:1713–1723

Alm Bergvall U, Rautio P, Kesti K et al (2006) Associational effects of plant defences in relation to within- and between-patch food choice by a mammalian herbivore: neighbour contrast susceptibility and defence. Oecologia 147:253–260

Ayal Y (2007) Trophic structure and the role of predation in shaping hot desert communities. J Arid Environ 68:171–187

Barbosa P, Hines J, Kaplan I et al (2009) Associational resistance and associational susceptibility: having right or wrong neighbors. Annu Rev Ecol Evol Syst 40:1–20

Barnes AD, Scherber C, Brose U et al (2020) Biodiversity enhances the multitrophic control of arthropod herbivory. Sci Adv 6:eabb6603

Bar-On YM, Phillips R, Milo R (2018) The biomass distribution on earth. Proc Natl Acad Sci 115:6506–6511

Bengtsson J, Bullock JM, Egoh B et al (2019) Grasslands – more important for ecosystem services than you might think. Ecosphere 10:e02582

Bernays EA, Bright KL (1993) Mechanisms of dietary mixing in grasshoppers: a review. Comp Biochem Physiol A Physiol 104:125–131

Bigger DS, Marvier MA (1998) How different would a world without herbivory be? A search for generality in ecology. Integr Biol Issues News Rev 1:60–67

Borer ET, Seabloom EW, Gruner DS et al (2014) Herbivores and nutrients control grassland plant diversity via light limitation. Nature 508:517–520

Bröcher M (2025) Plant-consumer interactions as cause and consequence of long-term BEF relationships. PhD thesis,. Friedrich-Schiller-Universität Jena

Bröcher M, Ebeling A, Hertzog L et al (2023) Effects of plant diversity on species-specific herbivory: patterns and mechanisms. Oecologia 201:1053–1066

Bröcher M, Meyer ST, Eisenhauer N, Ebeling A (2024) The positive plant diversity/consumer relationship is independent of grassland age. Basic Appl Ecol 76:58–68

Bröcher M, Ebeling A, Bassi L et al (2025) Plant–herbivore interactions depend on plant richness, plant and soil history. Funct Ecol 0:1–16

Cappelli SL, Pichon NA, Mannall T, Allan E (2022) Partitioning the effects of plant diversity on ecosystem functions at different trophic levels. Ecol Monogr 92:e1521

Cardinale BJ, Matulich KL, Hooper DU et al (2011) The functional role of producer diversity in ecosystems. Am J Bot 98:572–592

Carmona D, Lajeunesse MJ, Johnson MTJ (2011) Plant traits that predict resistance to herbivores. Funct Ecol 25:358–367

Coverdale TC, Davies AB (2023) Unravelling the relationship between plant diversity and vegetation structural complexity: a review and theoretical framework. J Ecol 111:1378–1395

Croijmans L, Mertens D, van Apeldoorn DF et al (2025) Herbivore prevalence poorly predicts yield in diverse cropping systems. J Appl Ecol 62:1229–1242

Dinnage R (2013) Phylogenetic diversity of plants alters the effect of species richness on invertebrate herbivory. PeerJ 1:e93

Ebeling A, Meyer ST, Abbas M et al (2014) Plant diversity impacts decomposition and herbivory via changes in aboveground arthropods. PLoS One 9:e106529

Ebeling A, Hines J, Hertzog LR et al (2018a) Plant diversity effects on arthropods and arthropod-dependent ecosystem functions in a biodiversity experiment. Basic Appl Ecol 26:50–63

Ebeling A, Rzanny M, Lange M et al (2018b) Plant diversity induces shifts in the functional structure and diversity across trophic levels. Oikos 127:208–219

Egorov E, Gossner MM, Meyer ST et al (2017) Does plant phylogenetic diversity increase invertebrate herbivory in managed grasslands? Basic Appl Ecol 20:40–50

Eisenhauer N, Mueller K, Ebeling A et al (2024) The multiple-mechanisms hypothesis of biodiversity–stability relationships. Basic Appl Ecol 79:153–166

Eskelinen A, Harpole WS, Jessen M-T et al (2022) Light competition drives herbivore and nutrient effects on plant diversity. Nature 611:301–305

Fischer M, Weyand A, Rudmann-Maurer K, Stöcklin J (2012) Omnipresence of leaf herbivory by invertebrates and leaf infection by fungal pathogens in agriculturally used grasslands of the Swiss Alps, but low plant damage. Alp Bot 122:95–107

Fricke U, Redlich S, Zhang J et al (2022) Plant richness, land use and temperature differently shape invertebrate leaf-chewing herbivory on plant functional groups. Oecologia 199:407–417

Gibson DJ (2009) Grasses and grassland ecology. Oxford University Press, Oxford

Gómez JM, González-Megías A (2007) Long-term effects of ungulates on phytophagous insects. Ecol Entomol 32:229–234

Gong B, Zhang G (2014) Interactions between plants and herbivores: a review of plant defense. Acta Ecol Sin 34:325–336

Gossner MM, Weisser WW, Meyer ST (2014) Invertebrate herbivory decreases along a gradient of increasing land-use intensity in German grasslands. Basic Appl Ecol 15:347–352

Grimaldi D, Engel MS (2005) Evolution of the insects. Cambridge University Press, Cambridge

Guo X, Petermann JS, Schittko C, Wurst S (2016) Root-knot nematodes (*Meloidogyne hapla*) can modify the positive plant intraspecific diversity-productivity effect on red clover in clover-grass communities. Appl Soil Ecol 102:26–35

Gutiérrez Y, Ott D, Scherber C (2020) Direct and indirect effects of plant diversity and phenoxy herbicide application on the development and reproduction of a polyphagous herbivore. Sci Rep 10:7300

Habel JC, Dengler J, Janišová M et al (2013) European grassland ecosystems: threatened hotspots of biodiversity. Biodivers Conserv 22:2131–2138

Haddad NM, Tilman D, Haarstad J et al (2001) Contrasting effects of plant richness and composition on insect communities: a field experiment. Am Nat 158:17–35

Halliday FW, Cappelli SL, Laine A-L (2023, 2023) Multiple dimensions of biodiversity mediate effects of temperature on invertebrate herbivory in a montane grassland. Oikos:e10028

Halliday FW, Everingham SE, Bröcher M et al (2025) Towards an integrative mechanistic framework for biodiversity–consumer relationships. Trends Ecol Evol 40:539–553

Hamann E, Blevins C, Franks SJ et al (2021) Climate change alters plant–herbivore interactions. New Phytol 229:1894–1910

Hanley ME (2004) Seedling herbivory and the influence of plant species richness in seedling neighbourhoods. Plant Ecol 170:35–41

Hanley ME, Lamont BB, Fairbanks MM, Rafferty CM (2007) Plant structural traits and their role in anti-herbivore defence. Perspect Plant Ecol Evol Syst 8:157–178

Hobohm C, Bruchmann I (2009) Endemische Gefäßpflanzen und ihre Habitate in Europa–Plädoyer für den Schutz der Grasland-Ökosysteme. Berichte Reinhold-Tüxen-Ges 21:142–161

Holt RD, Barfield M (2013) Direct plant–predator interactions as determinants of food chain dynamics. J Theor Biol 339:47–57

Hudewenz A, Klein A-M, Scherber C et al (2012) Herbivore and pollinator responses to grassland management intensity along experimental changes in plant species richness. Biol Conserv 150:42–52

Hülber K, Haider JA, Hager TE et al (2015) Insect herbivory in alpine grasslands is constrained by community and host traits. J Veg Sci 26:663–673

Huntzinger M, Karban R, Cushman JH (2008) Negative effects of vertebrate herbivores on invertebrates in a coastal dune community. Ecology 89:1972–1980

Hutchinson GE (1959) Homage to Santa Rosalia or why are there so many kinds of animals? Am Nat 93:145–159

Kigathi RN, Weisser WW, Reichelt M et al (2019) Plant volatile emission depends on the species composition of the neighboring plant community. BMC Plant Biol 19:58

Kluth S, Kruess A, Tscharntke T (2002) Insects as vectors of plant pathogens: mutualistic and antagonistic interactions. Oecologia 133:193–199

Korell L, Schmidt R, Bruelheide H et al (2016) Mechanisms driving diversity–productivity relationships differ between exotic and native communities and are affected by gastropod herbivory. Oecologia 180:1025–1036

Kozlov MV, Zvereva EL (2017) Background insect herbivory: impacts, patterns and methodology. In: Progress in botany. Springer, pp 313–355

Künast R, Weisser WW, Seibold S et al (2025) Differential effect of grassland mowing on arthropod taxa. Ecol Entomol 50:288–298

Lal R (2004) Soil carbon sequestration impacts on global climate change and food security. Science 304:1623–1627

Lanta V (2007) Effect of slug grazing on biomass production of a plant community during a short-term biodiversity experiment. Acta Oecol 32:145–151

Lau JA, Strengbom J, Stone LR et al (2008) Direct and indirect effects of CO2, nitrogen, and community diversity on plant–enemy interactions. Ecology 89:226–236

Li Y, Schuldt A, Ebeling A et al (2024) Plant diversity enhances ecosystem multifunctionality via multitrophic diversity. Nat Ecol Evol:1–11

Lin P-A, Kansman J, Chuang W-P et al (2023) Water availability and plant–herbivore interactions. J Exp Bot 74:2811–2828

Lipowsky A, Schmid B, Roscher C (2011) Selection for monoculture and mixture genotypes in a biodiversity experiment. Basic Appl Ecol 12:360–371

Loranger J, Meyer ST, Shipley B et al (2012) Predicting invertebrate herbivory from plant traits: evidence from 51 grassland species in experimental monocultures. Ecology 93:2674–2682

Loranger J, Meyer ST, Shipley B et al (2013) Predicting invertebrate herbivory from plant traits: polycultures show strong nonadditive effects. Ecology 94:1499–1509

Loranger H, Weisser WW, Ebeling A et al (2014) Invertebrate herbivory increases along an experimental gradient of grassland plant diversity. Oecologia 174:183–193

MacArthur RH (1972) Geographical ecology: patterns in the distribution of species. Princeton University Press

McArt SH, Halitschke R, Salminen J-P, Thaler JS (2013) Leaf herbivory increases plant fitness via induced resistance to seed predators. Ecology 94:966–975

Meyer ST, Ebeling A, Eisenhauer N et al (2016) Effects of biodiversity strengthen over time as ecosystem functioning declines at low and increases at high biodiversity. Ecosphere 7:e01619

Meyer ST, Scheithe L, Hertzog L et al (2017) Consistent increase in herbivory along two experimental plant diversity gradients over multiple years. Ecosphere 8:e01876

Mraja A, Unsicker SB, Reichelt M et al (2011) Plant community diversity influences allocation to direct chemical defence in *Plantago lanceolata*. PLoS One 6:e28055

Mulder CPH, Koricheva J, Huss-Danell K et al (1999) Insects affect relationships between plant species richness and ecosystem processes. Ecol Lett 2:237–246

Nitschke N, Ebeling A, Rottstock T et al (2010) Time course of plant diversity effects on *Centaurea jacea* establishment and the role of competition and herbivory. J Plant Ecol 3:109–121

Nitschke N, Wiesner K, Hilke I et al (2015) Increase of fast nutrient cycling in grassland microcosms through insect herbivory depends on plant functional composition and species diversity. Oikos 124:161–173

Nitschke N, Allan E, Zwölfer H et al (2017) Plant diversity has contrasting effects on herbivore and parasitoid abundance in *Centaurea jacea* flower heads. Ecol Evol 7:9319–9332

O'Mara FP (2012) The role of grasslands in food security and climate change. Ann Bot 110:1263–1270

Otway SJ, Hector A, Lawton JH (2005) Resource dilution effects on specialist insect herbivores in a grassland biodiversity experiment. J Anim Ecol 74:234–240

Parente L, Sloat L, Mesquita V et al (2024) Annual 30-m maps of global grassland class and extent (2000–2022) based on spatiotemporal machine learning. Sci Data 11:1303

Petermann JS, Buzhdygan OY (2021) Grassland biodiversity. Curr Biol 31:R1195–R1201

Petermann JS, Müller CB, Weigelt A et al (2010) Effect of plant species loss on aphid–parasitoid communities. J Anim Ecol 79:709–720

Pfisterer AB, Diemer M, Schmid B (2003) Dietary shift and lowered biomass gain of a generalist herbivore in species-poor experimental plant communities. Oecologia 135:234–241
Price PW, Denno RF, Eubanks MD et al (2011) Insect ecology: behavior, populations and communities. Cambridge University Press, Cambridge
Prieur-Richard A-H, Lavorel S, Linhart YB, Dos Santos A (2002) Plant diversity, herbivory and resistance of a plant community to invasion in Mediterranean annual communities. Oecologia 130:96–104
Reidinger S, Eschen R, Gange AC et al (2012) Arbuscular mycorrhizal colonization, plant chemistry, and aboveground herbivory on *Senecio jacobaea*. Acta Oecol 38:8–16
Ripple W, Newsome T, Wolf C et al (2015) Collapse of the world's largest herbivores. Sci Adv 1
Ristok C, Weinhold A, Ciobanu M et al (2023) Plant diversity effects on herbivory are related to soil biodiversity and plant chemistry. J Ecol 111:412–427
Root RB (1973) Organization of a plant-arthropod association in simple and diverse habitats: the fauna of collards (brassica Oleracea). Ecol Monogr 43:95–124
Roscher C, Schumacher J, Gubsch M et al (2012) Using plant functional traits to explain diversity–productivity relationships. PLoS One 7:e36760
Rossetti MR, Rösch V, Videla M et al (2019) Insect and plant traits drive local and landscape effects on herbivory in grassland fragments. Ecosphere 10:e02717
Russell EP (1989) Enemies hypothesis: a review of the effect of vegetational diversity on predatory insects and parasitoids. Environ Entomol 18:590–599
Scherber C, Milcu A, Partsch S et al (2006a) The effects of plant diversity and insect herbivory on performance of individual plant species in experimental grassland. J Ecol 94:922–931
Scherber C, Mwangi PN, Temperton VM et al (2006b) Effects of plant diversity on invertebrate herbivory in experimental grassland. Oecologia 147:489–500
Scherber C, Heimann J, Köhler G et al (2010) Functional identity versus species richness: herbivory resistance in plant communities. Oecologia 163:707–717
Schmitz OJ, Hambäck PA, Beckerman AP (2000) Trophic cascades in terrestrial systems: a review of the effects of carnivore removals on plants. Am Nat 155:141–153
Seabloom EW, Kinkel L, Borer ET et al (2017) Food webs obscure the strength of plant diversity effects on primary productivity. Ecol Lett 20:505–512
Shinohara N, Uchida K, Yoshida T (2019) Contrasting effects of land-use changes on herbivory and pollination networks. Ecol Evol 9:13585–13595
Silvertown J, Poulton P, Johnston E et al (2006) The park grass experiment 1856–2006: its contribution to ecology. J Ecol 94:801–814
Snyder WE, Snyder GB, Finke DL, Straub CS (2006) Predator biodiversity strengthens herbivore suppression. Ecol Lett 9:789–796
Soussana J-F, Allard V, Pilegaard K et al (2007) Full accounting of the greenhouse gas (CO2, N2O, CH4) budget of nine European grassland sites. Agric Ecosyst Environ 121:121–134
Srivastava DS, Lawton JH (1998) Why more productive sites have more species: an experimental test of theory using tree-hole communities. Am Nat 152:510–529
Stein C, Unsicker SS, Kahmen A et al (2010) Impact of invertebrate herbivory in grasslands depends on plant species diversity. Ecology 91:1639–1650
Terborgh JW (2015) Toward a trophic theory of species diversity. Proc Natl Acad Sci 112:11415–11422
Těšitel J, Tahadlová M, Lepš J, Hölzel N (2021) Linking insect herbivory with plant traits: phylogenetically structured trait syndromes matter. J Veg Sci 32
Unsicker SB, Baer N, Kahmen A et al (2006) Invertebrate herbivory along a gradient of plant species diversity in extensively managed grasslands. Oecologia 150:233–246
WallisDeVries MF, Van Swaay CA (2009) Grasslands as habitats for butterflies in Europe. Grassl Eur High Nat Value KNNV Publ Zeist, pp 27–34
White RP, Murray S, Rohweder M et al (2000) Grassland ecosystems. World Resources Institute
Wilson SD (1994) The contribution of grazing to plant diversity in alpine grassland and heath. Aust J Ecol 19:137–140
Yang Y, Joern A (1994) Compensatory feeding in response to variable food quality by *Melanoplus differentialis*. Physiol Entomol 19:75–82

4 Plant Diversity Phylogenetically and Functionally Structures Herbivore Communities and Plant-Herbivore Interactions: From Snapshot to Temporal Perspectives

Ming-Qiang Wang, Jing-Ting Chen, Rui Cheng, Arong Luo, and Chao-Dong Zhu

Abstract

Taxonomic, phylogenetic, and functional diversity of plants can shape multiple components of herbivore communities and plant-herbivore networks and in so doing exert controls over ecosystems. While a vast number of experimental plant diversity studies have offered valuable insights into effects on plant-herbivore interactions from a snapshot perspective, there is now a growing focus on the temporal unfolding of these effects and their implications for the stability of ecosystem functions and long-term effects on plants and associated fauna, especially in the context of the current global biodiversity crisis and pest outbreaks. Recent research highlights that temporal dynamics are crucial in shaping herbivore communities and plant-herbivore interactions: Diverse plant communities stabilize herbivore populations and communities through asynchronous resource availability and host-use complementarity, buffering temporal fluctuations in feeding pressure. Plant growth and phenological asynchrony, together with functional dissimilarities among herbivores, play key roles in bottom-up regulatory processes, influencing how resources and interactions propagate across trophic levels. Such temporal mechanisms underscore that biodiversity not only determines the composition of ecological communities but also their stability through

M.-Q. Wang (✉)
Mountain Ecological Restoration and Biodiversity Conservation Key Laboratory of Sichuan Province, Chengdu Institute of Biology, Chinese Academy of Sciences, Chengdu, China
e-mail: wangmq@cib.ac.cn

J.-T. Chen · R. Cheng · A. Luo · C.-D. Zhu
State Key Laboratory of Animal Biodiversity Conservation and Integrated Pest Management, Institute of Zoology, Chinese Academy of Sciences, Beijing, China

X. Moreira et al. (eds.), *Effects of Plant Diversity on Plant-Herbivore Interactions*, Ecological Studies 255, https://doi.org/10.1007/978-3-032-26461-9_4

time. Future studies are needed to better understand these temporal dynamics and associated long-term effects on plant-herbivore interactions and ecosystem functioning.

4.1 Introduction

Plant–herbivore interactions fundamentally shape ecosystem structure and function across biomes. Herbivory influences plant population dynamics and community composition (Barnes et al. 2020; Li et al. 2023; Wang et al. 2025a) and contributes to ecosystem services and disservices, including pollination, nutrient cycling, and pest outbreaks (Getman-Pickering et al. 2023; Popkin 2021; Pureswaran et al. 2018). In recent decades, ecological research has increasingly emphasized that the consequences of biodiversity loss cannot be fully understood without considering altered linkages between plants and herbivores (Eisenhauer et al. 2019; Schuldt et al. 2018; Srednick and Swearer 2024; Voglar et al. 2019). Within this context, plant diversity has been shown to play a central role in structuring herbivore assemblages and herbivore pressure (Haase et al. 2015; Haddad et al. 2011; Jactel et al. 2021; Wang et al. 2019; Wetzel et al. 2023), with implications for community stability and ecosystem resilience (Alavez et al. 2023).

Early plant diversity studies documenting effects on plant–herbivore relationships were largely based on manipulations of plant species richness (i.e., taxonomical) diversity, with evidence for positive effects herbivore richness via increases in resource availability and niche opportunities (Scherber et al. 2010a). However, evolutionary relatedness and plant functional traits—such as leaf economic strategies and chemical defence syndromes—strongly determine host suitability and consumer specialization (Díaz et al. 2013; Terhorst et al. 2018; Wang et al. 2020). Accordingly, phylogenetically diverse plant communities host more specialized herbivores, because closely related plants often share defensive traits and herbivore guilds (Pearse and Hipp 2009), while functionally diverse plant assemblages can enhance resource complementarity, potentially reducing pest outbreak risk (Wang et al. 2025a; Wetzel et al. 2023). For example, variation in leaf chemical and structural defences across functionally diverse tree mixtures can alter herbivore feeding niches and sustain more balanced herbivore communities (Castagneyrol et al. 2014a; Jorge et al. 2017; Schuldt et al. 2019). Overall, plant species richness represents variation in plant phylogenetic and functional diversity, which shapes herbivore communities and, in turn, influences both plant community and broader ecosystem services and disservices (Fig. 4.1).

Despite these significant advances, most empirical evidence still stems from snapshot studies capturing interactions at a single point in time, potentially overlooking the inherently dynamic nature of plant–herbivore relationships. Ecological interactions are dynamic, fluctuating across spatial and temporal scales. Variability in herbivory rates and host-use patterns are not the exception and can arise from seasonal phenology, interannual climate variation, and population asynchrony among species (van Klink et al. 2019; Wetzel et al. 2023). Recognizing these

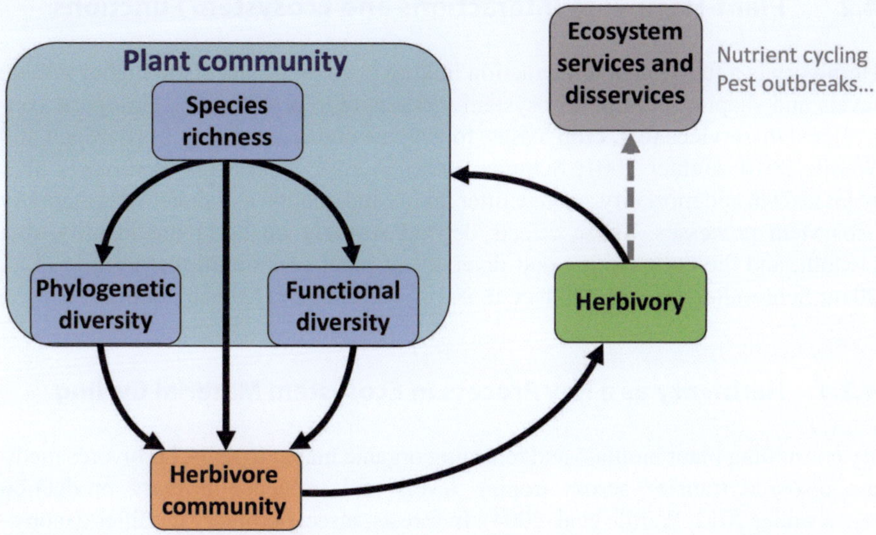

Fig. 4.1 Current conceptual framework of plant–herbivore community studies. Black solid lines denote direct effects, while grey dashed lines indicate potential indirect pathways mediated through intermediate processes

temporal dynamics is essential for understanding how biodiversity contributes to ecological stability through time via plant-herbivore interactions. Theoretical and empirical studies increasingly underscore that plant community asynchrony, herbivore functional dissimilarities, and temporal turnover in species interactions jointly buffer herbivore communities against environmental fluctuations (Aguilar et al. 2025; Haddad et al. 2011; Isbell et al. 2015; Mohammed 2024; van Klink et al. 2019; Wang et al. 2025a). Such findings suggest that plant diversity effects emerge not only from the number and identity of species present but also from the timing and variability in strength in plant-consumer interactions.

Building on these insights, this chapter synthesizes current knowledge on how phylogenetic and functional dimensions of plant diversity structure herbivore communities and mediate temporal variation in plant–herbivore interactions. Specifically, we aim to (1) review the contributions of plant–herbivore interactions to ecosystem functioning; (2) examine how plant taxonomic, phylogenetic, and functional diversity shape herbivore communities; (3) explore mechanisms and processes structuring plant–herbivore networks, including bottom-up and top-down effects; (4) examine how temporal variation in plant and herbivore dynamics influences community structure, stability, and resilience; and finally, (5) discuss how these temporal perspectives can improve our understanding and prediction of biodiversity loss and ecosystem responses under ongoing global change and altered pest pressure. Overall, by integrating trait-based, evolutionary, and temporal perspectives, this chapter aims to provide a conceptual framework for predicting how plant diversity loss may alter herbivore stability, ecosystem resilience, and the functioning of multitrophic systems in a changing world.

4.2 Plant-Herbivore Interactions and Ecosystem Functions

Herbivory is a fundamental interaction linking primary producers to higher trophic levels and shaping multiple ecosystem services (e.g. productivity, biological control) and disservices (e.g. crop losses, forest pest outbreaks, Fig. 4.2) (Bardgett and Wardle 2010; Hunter 2001; Schmitz 2008). Across ecosystems, herbivores alter plant growth and mortality, change litter inputs and quality, and thereby regulate key ecosystem processes. These effects depend strongly on herbivore identity, diet breadth, and the composition and diversity of plant communities (Metcalfe et al. 2014; Schowalter 2012; Schuldt et al. 2010; Thébault and Loreau 2006).

4.2.1 Herbivory as a Key Process in Ecosystem Material Cycling

By consuming plant biomass and returning organic matter to soils, herbivores mediate resource transfer across trophic levels and influence primary production (Schowalter 2012; Wardle et al. 2004). In forests, insect herbivory modifies canopy–soil linkages by altering leaf chemistry, throughfall composition, and litter quality, which in turn influences microbial activity and soil nutrient availability (Belovsky and Slade 2000; Kristensen et al. 2020; Metcalfe et al. 2014). Frass stoichiometry—reflecting herbivore diet and assimilation efficiency—can increase soil N and P fluxes and stimulate microbial mineralization, feeding back to plant nutrient status and regrowth (Cherif and Loreau 2013; Frost and Hunter 2004; Kagata and Ohgushi

Fig. 4.2 Pest outbreak caused by the European spruce bark beetle (*Ips typographus*) in spruce forests of Harz National Park (Germany). (Photo credit: Ming-Qiang Wang)

2012). Periodic defoliation also create temporal pulses of nutrient release, followed by compensatory plant growth (Hättenschwiler and Gasser 2005; Hunter 2001). Collectively, these mechanisms position herbivory as both a top-down regulator of plant biomass and a bottom-up driver of nutrient renewal and ecosystem productivity.

4.2.2 Roles of Plant–Herbivore Interactions in Regulating Ecosystem Services and Disservices

Plant–herbivore interactions generate ecosystem services and disservices primarily through variation in the magnitude and distribution of herbivore pressure across plant communities. Differences in herbivore load, host susceptibility, and feeding compensatory capacity determine whether biomass removal stabilizes ecosystem functioning or leads to pronounced degradation (Bardgett and Wardle 2010; Hunter 2001; Schmitz 2008). At low to moderate intensities, herbivory can contribute to regulating services by limiting plant dominance, maintaining vegetation turnover, and sustaining ecosystem productivity (Visakorpi et al. 2024). At high intensities, however, herbivory constitutes a major ecosystem disservice by reducing primary production, impairing resource availability, and triggering large economic and ecological losses (Hwang et al. 2024; lM-Arnold et al. 2016; Schaefer et al. 2010). Globally, herbivore damage represents one of the most pervasive biotic constraints on plant-based production systems, with consequences that propagate to food security, carbon storage, and ecosystem health (Deutsch et al. 2018; Oerke 2006; Tanentzap and Coomes 2012). Importantly, the balance between services and disservices depends on the magnitude of herbivore pressure, the susceptibility of host plants, and the capacity of ecosystems to absorb or compensate for biomass loss (Osterrieth and Bosker 2024; Scherber et al. 2010b). These outcomes emerge from direct plant–herbivore interactions rather than requiring additional trophic complexity, highlighting herbivory itself as a core ecological process governing ecosystem performance across biomes (Hunter et al. 2012; Pinho et al. 2018).

4.2.3 Examples Across Ecosystems: Forests, Grasslands, and Agroecosystems

4.2.3.1 Forests

Forest ecosystems provide clear examples of two-way coupling between plant community structure, ecosystem functioning, and herbivore-mediated dynamics. Large tree-diversity experiments have repeatedly shown that increasing tree species richness alters herbivore damage (Schuldt et al. 2010), consumer richness (Guo et al. 2021; Schuldt et al. 2019; Wang et al. 2019), and the architecture of multitrophic interactions as structured around plant–herbivore links (Staab et al. 2015; Staab et al. 2016; Wang et al. 2025b; Wang et al. 2025c; Wang et al. 2020) and that these changes in turn influence productivity, carbon storage, and soil processes (Li et al. 2024; Schuldt et al. 2018; Tang et al. 2022). Importantly, most of these ecosystem-level consequences arise from how plant diversity restructures herbivore community composition, feeding pressure, and host-use patterns rather than from direct effects on predators. For example, early studies found positive effects of plant species richness on herbivory at some scales, but the net consequences for productivity were mediated by plant functional trait composition and herbivore niche differentiation: Higher functional diversity increased resource complementarity for some generalist herbivores while also diluting host availability for specialists, producing complex indirect effects on productivity (Li et al. 2023; Tang et al. 2022). Other analyses revealed scale-dependent turnover in Lepidoptera communities (Wang et al. 2023) and showed that tree growth, microclimate, and tree compositional heterogeneity independently structure herbivore community stability (Wang et al. 2025a).

Mechanistically, in mixed-tree species stands, host dilution and increased plant apparency heterogeneity reduce host-finding efficiency and feeding success of specialist herbivores, while resource heterogeneity sustains higher overall herbivore richness (Castagneyrol et al. 2013; Jactel and Brockerhoff 2007; Shen et al. 2025). Together, these bottom-up effects primarily reshape herbivore abundance distributions and damage allocation across host species, constituting the main pathway through which plant diversity regulates forest functioning in this context. Increased structural complexity associated with higher plant diversity may also indirectly influence predator and parasitoid activity, but these effects primarily remain secondary to herbivore-driven pathways in regulating ecosystem processes (Hunter 2001; Schmitz 2008; Wang et al. 2025c). Microclimatic buffering under diverse canopies further modifies herbivore development rates and feeding intensity, thereby reinforcing diversity–herbivory–function feedbacks (Schnabel et al. 2025; Schuldt et al. 2019). Conversely, monocultures or low-diversity stands provide homogeneous, high-density host resources that favour specialist outbreaks or facilitate rapid population growth of a few pest species, often producing strong negative impacts on carbon sequestration and timber quality (Jactel and Brockerhoff 2007; Seidl et al. 2017).

4.2.3.2 Grasslands

Grassland ecosystems provide a well-established model for how plant community structure and resource heterogeneity regulate herbivore pressure while also allowing consumer-driven feedbacks to emerge under certain conditions. Large mammalian grazers rework vegetation structure and nutrient distributions by selective foraging, trampling, and dung/urine deposition, producing nutrient "hotspots" and grazing lawns that alter plant species composition and productivity (Knapp et al. 1999; McNaughton 1984). These bottom-up changes in forage quality and spatial structure often precede and condition the strength of subsequent grazer feedbacks, which can reinforce particular vegetation states and stabilize grazer–plant interactions (Augustine and McNaughton 2006). Importantly, these consumer-driven modifications of plant community and soil resources feedback to herbivore demography—for example, by enhancing juvenile body growth rates (e.g. weight gain or development speed), fecundity, and survival where forage quality is improved (Belovsky and Slade 2000; Belsky 1992; Johnson et al. 2021).

Insect herbivores in grasslands (e.g. grasshoppers, lepidopteran larvae) also exert measurable effects on nutrient cycling and plant competitive interactions (Belovsky and Slade 2000; Metcalfe et al. 2014). Experimental studies show that insect herbivory can accelerate decomposition and nitrogen mineralization via frass inputs and by changing both the quantity and quality of litter, thereby affecting subsequent plant regrowth and species turnover (Metcalfe et al. 2014). At the same time, plant traits, soil fertility, and climate regimes impose strong bottom-up constraints on insect body size, voltinism, and outbreak potential—so warming or altered precipitation can shift which herbivore functional groups dominate and the magnitude of their impacts (Hance et al. 2007; Prather et al. 2017). Overall, grasslands exhibit tightly coupled interactions in which plant-driven variation in resource quality regulates herbivore pressure, and herbivore activity in turn feeds back to vegetation and nutrient dynamics.

4.2.3.3 Agroecosystems

Agricultural landscapes provide clear evidence of how management-driven variation in plant diversity and resource heterogeneity structures herbivore communities and shapes their consequences for ecosystem services and disservices. Diversified cropping systems—such as intercropping, hedgerows, cover crops, and floral strips—create heterogeneous plant resource environments that disrupt host finding, dilute specialized pests, and promote more even herbivore assemblages (Chaplin-Kramer et al. 2011; Letourneau et al. 2011). These bottom-up changes in host quality and spatial structure often explain more variance in herbivore pressure than natural-enemy effects alone, even though diversified habitats may incidentally also enhance parasitoid and predator performance (Gurr et al. 2017; Tscharntke et al. 2005).

Conversely, intensive monocultures generate homogeneous, nutrient-enriched plant resources that increase herbivore growth rates, fecundity, and outbreak probability, especially when combined with fertilizer-driven increases in host nutritional value (Deutsch et al. 2018; Oerke 2006; White 2012). Landscape simplification further amplifies these bottom-up effects by reducing spatial and temporal variability in host plants, which facilitates pest dominance and increases outbreak probability (Gurr et al. 2017; Tscharntke et al. 2005). Thus, agroecosystems illustrate that management decisions altering plant diversity, host quality, and landscape heterogeneity exert strong bottom-up control over herbivore population dynamics and thereby mediate both ecosystem services (e.g., yield stability, pollination–pest interactions) and disservices (e.g. crop damage, outbreak risk).

4.3 Functional and Phylogenetic Sources of Plant Diversity Structuring Herbivore Communities and Underlying Mechanisms

Plant diversity—taxonomic, phylogenetic, and functional—strongly structures herbivore communities through multiple, partly overlapping mechanisms, including the filtering of herbivore assemblages via host availability and trait matching, as well as changes in resource quality, spatial heterogeneity, and interaction strengths. Among these, community-level filtering represents a dominant pathway by which plant diversity constrains herbivore establishment and host use, while additional processes such as niche partitioning and density-dependent responses further shape herbivore abundance and damage patterns. Through these combined pathways, plant diversity modulates not only herbivore richness and composition but also the organization and temporal stability of herbivore communities across ecosystems and spatial scales (Haddad et al. 2011; Scherber et al. 2010a; Schuldt et al. 2019). These mechanisms collectively represent bottom-up pathways through which plant diversity propagates to ecosystem processes via herbivore-mediated effects (see Fig. 4.3 for potential research framework and hypotheses).

4.3.1 Taxonomic Diversity: Plant Species Richness and Composition and Their Links to Herbivore Diversity

Across grasslands and forests, higher plant species richness is generally associated with greater herbivore richness and altered community composition, reflecting expanded resource niches and more heterogeneous microhabitats (Scherber et al. 2010a; Schuldt et al. 2019; Siemann et al. 1998; Wang et al. 2019). Classic experimental work at Cedar Creek and Jena showed that plant richness and identity jointly predict arthropod diversity across trophic levels, with strong bottom-up effects on herbivores and their natural enemies (Haddad et al. 2009; Scherber et al. 2010a). In tree-diversity experiments, multiple components of plant taxonomic diversity (richness, identity, composition) were also found to simultaneously structure herbivore assemblages, emphasizing that no single diversity axis is sufficient to explain consumer patterns (Schuldt et al. 2019; Vehvilainen et al. 2007; Zhang et al. 2017).

Species composition and dominant identities matter, because host plants differ in palatability, phenology, and chemical defences; thus, who is present can be as important as how many are present (Coley and Barone 1996; Schaefer and Ruxton 2011; Wang et al. 2023). Herbivore community context further determines how plant taxonomic diversity effects are expressed at the herbivore level, with plant mixtures differing in their net outcomes for specialist versus generalist consumers depending on neighbourhood composition and dominance structure (Barbosa et al. 2009; Castagneyrol et al. 2014b; Jactel and Brockerhoff 2007). Together, these results indicate that plant species richness expands the potential niche space for herbivores, whereas plant composition sorts which herbivore lineages and feeding guilds actually dominate local communities.

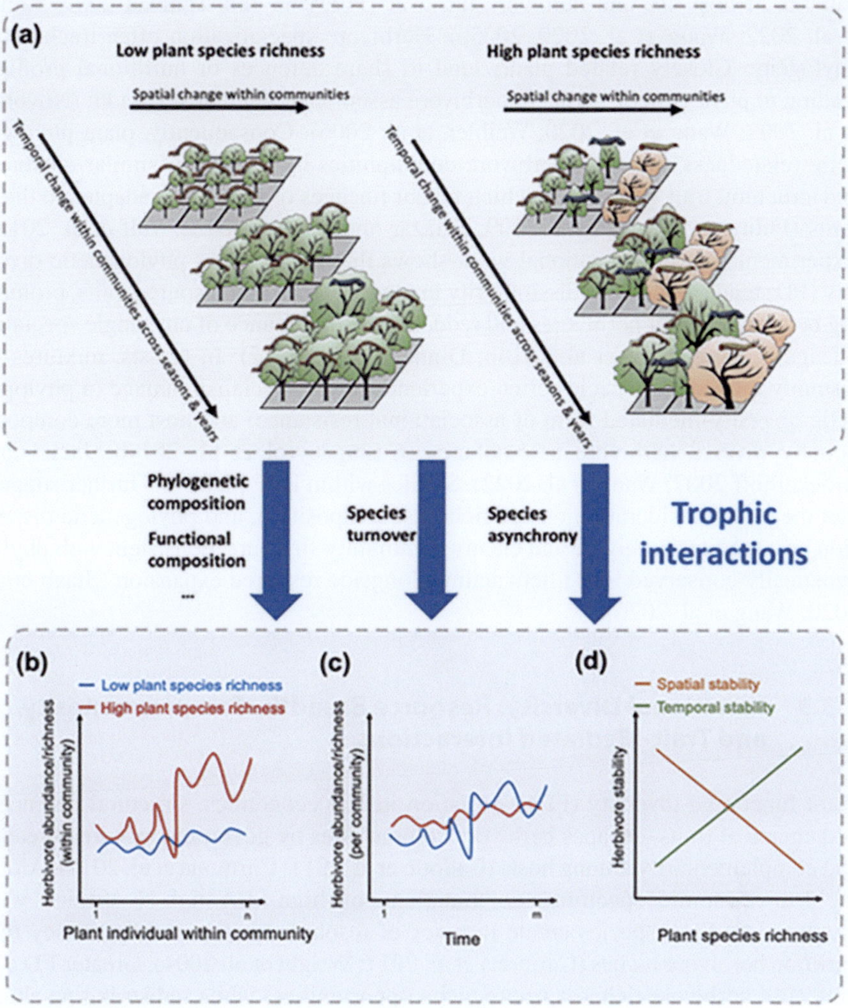

Fig. 4.3 A potential research framework on bottom-up effects of plant species richness across spatio-temporal scales. Specifically, plant community dynamics may drive herbivore community fluctuations via trophic interactions, mediated by phylogenetic and functional composition, species turnover, and species asynchrony. Herbivore might exhibit fluctuations within communities characterized across spatial (**b**) and temporal scales (**c**). Herbivore abundance and richness may exhibit higher temporal stability, driven by stable resource availability at the community level, but lower spatial stability due to greater inter- than intraspecific variation in host quality (**d**)

4.3.2 Phylogenetic Diversity: Conserved Traits, Host-Use Filtering, and the Role of Plant Relatedness

Because many plant traits relevant to herbivory are phylogenetically conserved (e.g., latex, cardenolides, tannins, leaf toughness), plant phylogeny can predict patterns of host use and herbivore turnover (Futuyma and Agrawal 2009; König et al. 2022; Wang et al. 2020, 2025a). Herbivore specialization often tracks host phylogeny: Closely related plants tend to share defences or nutritional profiles, leading to phylogenetic signal in herbivore assemblages and host breadth (Novotny et al. 2002; Wang et al. 2020; Weiblen et al. 2006). Consequently, plant phylogenetic relatedness can filter herbivore communities by clustering similar chemical and structural trait syndromes, which favour lineages of specialists adapted to those traits (Futuyma and Agrawal 2009; Salazar and Marquis 2022; Volf et al. 2018). Experimental and observational work shows that higher plant phylogenetic diversity (PD) tends to increase dissimilarity in plant defence and resource traits, promoting beta-diversity of herbivores and reducing the dominance of any single specialist lineage (Castagneyrol et al. 2014b; Dinnage et al. 2012). In forests, mixtures of distantly related tree species often experience lower specialist damage (a phylogenetic diversity-mediated form of associational resistance) and host more compositionally diverse herbivore assemblages (Castagneyrol et al. 2014b; Jactel and Brockerhoff 2007; Wang et al. 2022). Studies within BEF platforms further suggest that the joint consideration of tree richness, composition, and phylogenetic dispersion best predicts herbivore and enemy community structure, consistent with phylogenetically conserved trait filters acting alongside resource expansion (Staab et al. 2020; Wang et al. 2020).

4.3.3 Functional Diversity: Resource Breadth, Complementarity, and Trait-Mediated Interactions

Plant functional diversity (FD)—variation in leaf economics, structural defences, and chemical traits—shapes herbivore communities by governing resource breadth and complementarity among hosts (Cadotte et al. 2011; Carmona et al. 2011). Along the leaf-economic spectrum, combinations of high-SLA/high-N species with tougher, low-SLA species create mosaics of resource quality and apparency that partition herbivore niches (Carmona et al. 2011; Wright et al. 2004). Greater FD can thus raise herbivore richness (more niche opportunities) while redistributing abundance across guilds and decreasing per-species dominance (Scherber et al. 2010a; Schuldt et al. 2019; Wang et al. 2025c).

Crucially, trait syndromes (e.g. latex–cardenolide in milkweeds, condensed tannins–lignin in woody plants) mediate interaction strengths and tri-trophic linkages (Agrawal and Weber 2015; Ali and Agrawal 2012; Turlings and Erb 2018). Induced defences and volatile organic compounds can alter herbivore acceptance and performance while attracting parasitoids and predators, thereby reshaping network topology (Turlings and Erb 2018). Biodiversity experiments show that plant FD and trait composition predict herbivore feeding guild structure and natural enemy responses—signals that propagate to ecosystem processes such as herbivory rates and biological control (Scherber et al. 2010a; Schuldt et al. 2012). Integrating FD with PD clarifies mechanisms: Phylogeny captures shared ancestry of trait suites, while function quantifies the realized trait distances that directly interact with herbivore physiology and behaviour (Cadotte et al. 2011; Cadotte et al. 2010; Pearse and Hipp 2009).

4.3.4 Bottom-Up and Top-Down Mechanisms Shaping Plant Diversity–Herbivory Linkages

Plant–herbivore interactions are strongly conditioned by the resource environment provided by plants. Beyond determining food quantity, plant traits shape the nutritional quality and defensive landscape experienced by herbivores. Leaf traits such as nitrogen and water content can increase palatability and accelerate insect growth, whereas structural traits (e.g. toughness, trichomes) and chemical defences (e.g. phenolics, terpenoids) constrain feeding (Carmona et al. 2011; Muiruri et al. 2019; Pearse and Hipp 2012).

These trait-mediated effects generate predictable associations between plant lineages and herbivore communities, reinforcing the phylogenetic and functional diversity mechanisms described above. Legumes typically support higher herbivory due to elevated foliar nitrogen, whereas conifers often sustain lower damage owing to strong secondary metabolites (Endara and Coley 2011). Thus, bottom-up mechanisms provide the mechanistic basis through which taxonomic, phylogenetic, and functional diversity translate into herbivore community structure and ecosystem consequences.

While plant diversity primarily influences ecosystem functioning through herbivore-mediated bottom-up pathways, natural enemies can also modulate these dynamics under certain conditions. Predators and parasitoids suppress herbivore populations, potentially altering the magnitude and distribution of herbivore pressure (Mooney et al. 2012; Schmitz 2008). However, these top-down effects often depend on the same plant attributes that structure herbivores, including vegetation complexity and trait composition (Tylianakis et al. 2008). In this sense, enemy effects are typically indirect, operating through herbivore population regulation rather than constituting independent biodiversity–ecosystem function pathways.

4.3.5 Network Structure and Community Stability

Beyond pairwise interactions, the organization of plant–herbivore networks provides insights into how communities respond to perturbations. Structural features such as modularity, nestedness, and interaction specialization influence the robustness of interaction webs (Bascompte et al. 2003; Thebault and Fontaine 2010). High modularity can compartmentalize perturbations, limiting the spread of extinctions across the network, whereas nestedness enhances redundancy by ensuring that generalist species maintain connections even when specialists are lost (Gilarranz et al. 2017; Thebault and Fontaine 2010). The degree of interaction specialization further mediates community stability: While extreme specialization increases vulnerability to partner loss, moderate specialization can promote coexistence by reducing competitive overlap among herbivores (Blüthgen and Klein 2011).

Crucially, network structure reflects plant attributes. Functional and phylogenetic dissimilarities among plant hosts can promote greater modularity by supporting distinct herbivore assemblages, while high trait redundancy tends to generate more generalized network configurations (Giling et al. 2019; Rezende et al. 2009). Rather than representing separate mechanisms, these structural properties summarize how plant diversity shapes the distribution of interaction strengths within plant-herbivore communities.

4.4 Temporal Dynamics of Plant-Herbivore Interactions

4.4.1 From Snapshot to Temporal Scale

Most plant–herbivore research has relied on short-term observations or single time point experiments, yet these approaches neglect the temporal complexity of interactions which often holds the key for understanding interaction outcomes. Herbivory is not constant but fluctuates across seasons, years, and even decades in response to plant ontogeny, climate variation, and shifting consumer pressure (Hunter 2001). For instance, insect outbreaks in boreal and temperate forests may occur at multi-annual intervals, with cascading effects on nutrient cycling and forest regeneration that cannot be captured by short-term studies (Hagen et al. 2007; Jepsen et al. 2009; Peltonen et al. 2002; Popkin 2021). Similarly, grassland experiments show that the cumulative effects of grazing on soil fertility and community composition emerge only after years of repeated herbivore pressure (Knapp et al. 2012). Long-term biodiversity–ecosystem function experiments have also revealed that the strength and direction of stability–diversity relationships can change through time for herbivores and other trophic levels, underscoring the necessity of a temporal perspective (Gottschall et al. 2022; Wang et al. 2025a). As a whole, these findings highlight that static snapshots risk misleading conclusions, whereas temporal monitoring captures lagged feedbacks and changes in species turnover essential for understanding ecosystem functioning. Importantly, plant diversity can modify both the magnitude and predictability of these temporal fluctuations by altering resource heterogeneity, host turnover, and interaction redundancy across time (Haddad et al. 2011; Tilman et al. 2006; Wang et al. 2025a, 2026).

4.4.2 Drivers of Temporal Variation

Temporal dynamics of plant–herbivore interactions are shaped by both intra-annual processes and inter-annual variability. Within seasons, plant phenology determines when and for how long tissues are available and palatable. Differences in leaf flush, flowering, and recovery from damage generate temporal windows of susceptibility that herbivores can exploit (Feeny 1970; Turcotte et al. 2014). In diverse plant communities, asynchronous phenology stabilizes resource availability, buffering against synchronized defoliation and reducing the risk of pest outbreaks (van Asch and Visser 2007). Thus, higher plant diversity can dampen intra-annual peaks in herbivory pressure by spreading resource availability across time, thereby weakening synchronized consumer responses (Wang et al. 2025a). Herbivore phenology is equally critical: Mismatches between insect emergence and host plant development—exacerbated by climate warming—can reduce feeding efficiency, whereas prolonged overlaps between herbivore activity and host availability may intensify pressure within and across growing seasons (Kharouba et al. 2018; Singer and Parmesan 2010).

Across years, environmental variability and population feedbacks generate longer-term fluctuations. Episodic resource pulses, such as mast seeding or irregular leaf flushes, drive boom–bust dynamics in herbivore populations, while climatic fluctuations in temperature and precipitation alter both plant productivity and insect survival (Bogdziewicz et al. 2020; Parmesan and Hanley 2015; Schnabel et al. 2021). Cyclical defoliator outbreaks in forests reflect such inter-annual processes, arising from interactions among host recovery rates, herbivore demography, and natural enemy regulation (Haynes et al. 2014; Tenow et al. 2005). However, the extent to which such boom–bust cycles propagate through ecosystems depends strongly on the plant diversity context. In species-rich communities, lower host dominance and greater trait dispersion can dilute outbreak intensity by limiting the spatial and temporal continuity of preferred hosts (Jactel and Brockerhoff 2007), whereas low-diversity stands may amplify long-term cyclical dynamics through high host concentration (Root 1973). Functional differences among herbivores may further amplify these dynamics: Specialists often collapse in poor host years, whereas generalists persist by host switching, leading to guild restructuring over time (Lemoine and Burkepile 2012). Plant diversity mediates this restructuring by determining the availability of alternative hosts and the degree of resource complementarity across years. Long-term monitoring across Europe confirms these divergent trajectories, with specialist declines contrasting with the stability or even increase in abundance of generalist herbivores under global change (Engelhardt et al. 2022; Pilotto et al. 2020).

Together, these intra- and inter-annual drivers highlight that temporal variability in plant–herbivore interactions emerges from the interplay of plant growth asynchrony, herbivore life histories, and climatic fluctuations (Wang et al. 2025a). Crucially, plant diversity influences how these drivers translate into herbivory pressure and community turnover, thereby modulating the temporal variability of herbivory-dependent ecosystem processes such as primary production and nutrient cycling (Li et al. 2023).

4.4.3 Stability and Resilience of Herbivore Communities

A central question is how temporal variability translates into stability and resilience of herbivore communities and their ecosystem impacts. Biodiversity–stability theory predicts that higher species richness buffers ecosystems against fluctuations by providing compensatory dynamics among species (Isbell et al. 2015; Tilman and Downing 1994). Recent multitrophic experiments confirm that plant diversity enhances not only the temporal stability of plant biomass but also the stability of herbivore assemblages, in part by dampening the volatility of herbivore abundance across years (Huang et al. 2018; Wang et al. 2025a).

Beyond variance buffering, plant diversity can also influence other temporal properties of herbivory, including resistance (the magnitude of initial response to disturbance), recovery rates following perturbation, and the frequency and amplitude of herbivore outbreaks. Diverse plant communities often show reduced peak herbivory during extreme climatic events and faster post-disturbance recovery of both plant and herbivore assemblages, possibly reflecting greater interaction redundancy and resource complementarity (Liang et al. 2025; Wang et al. 2026).

Phylogenetic and functional diversity might further contribute by ensuring that herbivores partition host resources differently through time, reducing the likelihood of synchronized population crashes or outbreaks (Wang et al. 2019). By spreading herbivore impacts across functionally dissimilar hosts and temporal niches, plant diversity can prevent the concentration of damage on single dominant species, thereby stabilizing ecosystem productivity and nutrient retention (e.g. Hautier et al. 2014). Importantly, resilience is not only a function of diversity but also of interaction network structure: Modularity and redundancy within plant–herbivore networks promote recovery following disturbances such as droughts or pest outbreaks (Thebault and Fontaine 2010). In diverse systems, these structural properties are more likely to emerge, enhancing the capacity of herbivore communities to reorganize without triggering cascading ecosystem dysfunction. Accordingly, these long-term studies increasingly demonstrate that temporal stability in ecosystems is an emergent property of biodiversity across trophic levels rather than a simple outcome of plant or herbivore richness alone. Thus, plant diversity does not merely buffer variability; it actively shapes the temporal expression, intensity, and ecosystem consequences of herbivory across seasons and years.

4.5 Emerging Perspectives and Research Frontiers

Future research on plant–herbivore interactions should increasingly treat temporal variation as the central organizing axis around which phylogenetic, functional, and chemical dimensions of plant diversity are integrated. Rather than considering these perspectives separately, a key frontier lies in understanding how these sources of plant diversity independently and jointly modulate herbivore dynamics across time—shaping not only mean interaction strengths but also seasonal trajectories, inter-annual variability, disturbance responses, and long-term stability.

From a chemical perspective, incorporating plant volatile organic compounds (VOCs) within a temporal framework can reveal how diversity alters the timing, persistence, and synchrony of defence expression. Incorporating plant VOCs can provide mechanistic insights into how chemical defences mediate herbivore behaviour, plant–plant signalling, and ultimately community stability (Kessler et al. 2023; Medina-van et al. 2025). Recent work emphasizes that plants and herbivores are engaged in an information "arms race" in which plants evolve VOC-mediated defences and herbivores evolve mechanisms to detect or circumvent them (Zu et al. 2020). VOCs also mediate interactions among plants, influencing neighbour defence induction and community-level trait complementarity (Kigathi et al. 2019; Ling et al. 2022). However, these processes are inherently dynamic: VOC emissions fluctuate diurnally, seasonally, and under episodic stress, and plant diversity may alter both the magnitude and temporal predictability of these chemical signals. Diverse communities could desynchronize peak vulnerability by staggering induction timing across species, thereby reducing simultaneous herbivore attraction and dampening outbreak risk. Combining these axes allows testing whether evolutionary relatedness, trait complementarity, temporal asynchrony, and chemical interactions jointly constrain or promote community stability. Such integration is particularly timely given recent advances in long-term biodiversity experiments, phylogenomic resources, and chemical ecology methods, which allow simultaneous testing of evolutionary, functional, and chemical hypotheses across multiple scales (Cadotte et al. 2013; Kessler et al. 2023; Zu et al. 2020).

Another frontier lies in linking plant–herbivore interactions to ecosystem multifunctionality. While herbivory is often treated as a disservice due to its role in defoliation or pest outbreaks, it also regulates nutrient cycling, shapes successional trajectories, and indirectly affects other key processes such as pollination and decomposition (Schuldt et al. 2018; Soliveres et al. 2016; van der Plas 2019). Adopting a temporal lens highlights that plant diversity may influence not only the average contribution of herbivory to multifunctionality but also the timing, duration, and recovery of herbivore-driven effects on ecosystem processes. For instance, in species-rich systems, diffuse and temporally distributed herbivore pressure may promote gradual nutrient return and sustained decomposition dynamics, whereas concentrated outbreaks in low-diversity stands may trigger abrupt nutrient pulses and transient functional instability. Integrating multitrophic datasets is therefore essential to reveal when and how herbivore-driven processes reinforce or trade off with other ecosystem functions, with implications for management under global change.

At the same time, advances in ecological network theory provide opportunities to incorporate network dynamics into forecasting and management. Dynamic plant–herbivore networks, informed by temporal turnover and interaction rewiring, can act as early-warning signals of destabilization before dramatic biodiversity losses occur (Gómez et al. 2023; Neff et al. 2021; Tylianakis and Morris 2017). Critically, plant diversity may shape the tempo and mode of such network rewiring through time: High-diversity communities may exhibit gradual interaction turnover and compensatory rewiring, whereas simplified systems may show abrupt collapses in connectivity following disturbance. Embedding network metrics within long-term monitoring frameworks would allow testing whether diversity reduces the velocity or amplitude of temporal network change under climate extremes. Linking such network-based indicators with remote sensing or trait-based monitoring tools may enhance predictive models of outbreak risk and ecosystem resilience (Pellissier et al. 2018).

Finally, cross-biome comparisons and global syntheses are urgently needed to capture context dependency, particularly in how plant diversity modulates temporal variability across contrasting climatic regimes. Biome-specific drivers—such as drought in drylands, canopy stratification in tropical forests, or freeze–thaw cycles in boreal systems—may fundamentally alter the strength and direction of biodiversity–stability relationships. A temporally explicit comparative framework would help determine whether diversity consistently dampens inter-annual variability and outbreak amplitude or whether its stabilizing role weakens under certain disturbance regimes. Coordinated global networks and meta-analyses, such as the sPlot or NutNet initiatives, will be critical for disentangling general principles from regional contingencies (Bruelheide et al. 2019; Welti and Kaspari 2021). Together, these perspectives highlight that future research must adopt a truly integrative and comparative approach to understand and forecast the ecological and functional consequences of plant–herbivore interactions under temporally explicit perspectives.

References

Agrawal AA, Weber MG (2015) On the study of plant defence and herbivory using comparative approaches: how important are secondary plant compounds. Ecol Lett 18:985–991

Aguilar J, Suweis S, Maritan A, Azaele S (2025) Unraveling the temporal dependence of ecological interaction measures. arXiv preprint https://arxiv.org/abs/2508.19197

Alavez V, Santos-Gally R, Gutiérrez-Aguilar M, Del-Val E, Boege K (2023) Influence of phylogenetic diversity of plant communities on tri-trophic interactions. Oecologia 203:125–137

Ali JG, Agrawal AA (2012) Specialist versus generalist insect herbivores and plant defense. Trends Plant Sci 17:293–302

Augustine DJ, McNaughton SJ (2006) Interactive effects of ungulate herbivores, soil fertility, and variable rainfall on ecosystem processes in a semi-arid savanna. Ecosystems 9:1242–1256

Barbosa P, Hines J, Kaplan I, Martinson H, Szczepaniec A, Szendrei Z (2009) Associational resistance and associational susceptibility: having right or wrong neighbors. Annu Rev Ecol Evol Syst 40:1–20

Bardgett RD, Wardle DA (2010) Aboveground-belowground linkages: biotic interactions, ecosystem processes, and global change. Oxford University Press

Barnes AD, Scherber C, Brose U et al (2020) Biodiversity enhances the multitrophic control of arthropod herbivory. Sci Adv 6. https://doi.org/10.1126/sciadv.abb6603

Bascompte J, Jordano P, Melián CJ, Olesen JM (2003) The nested assembly of plant–animal mutualistic networks. Proc Natl Acad Sci 100:9383–9387

Belovsky GE, Slade JB (2000) Insect herbivory accelerates nutrient cycling and increases plant production. Proc Natl Acad Sci USA 97:14412–14417. https://doi.org/10.1073/pnas.250483797

Belsky AJ (1992) Effects of grazing, competition, disturbance and fire on species composition and diversity in grassland communities. J Veg Sci 3:187–200

Blüthgen N, Klein A-M (2011) Functional complementarity and specialisation: the role of biodiversity in plant–pollinator interactions. Basic Appl Ecol 12:282–291. https://doi.org/10.1016/j.baae.2010.11.001

Bogdziewicz M, Canelo T, Bonal R (2020) Rainfall and host reproduction regulate population dynamics of a specialist seed predator. Ecol Entomol 45:26–35

Bruelheide H, Dengler J, Jiménez-Alfaro B et al (2019) sPlot–A new tool for global vegetation analyses. J Veg Sci 30:161–186

Cadotte MW, Jonathan DT, Regetz J, Kembel SW, Cleland E, Oakley TH (2010) Phylogenetic diversity metrics for ecological communities: integrating species richness, abundance and evolutionary history. Ecol Lett 13:96–105. https://doi.org/10.1111/j.1461-0248.2009.01405.x

Cadotte MW, Carscadden K, Mirotchnick N (2011) Beyond species: functional diversity and the maintenance of ecological processes and services. J Appl Ecol 48:1079–1087. https://doi.org/10.1111/j.1365-2664.2011.02048.x

Cadotte M, Albert CH, Walker SC (2013) The ecology of differences: assessing community assembly with trait and evolutionary distances. Ecol Lett 16:1234–1244. https://doi.org/10.1111/ele.12161

Carmona D, Lajeunesse MJ, Johnson MTJ (2011) Plant traits that predict resistance to herbivores. Funct Ecol 25:358–367. https://doi.org/10.1111/j.1365-2435.2010.01794.x

Castagneyrol B, Giffard B, Pere C, Jactel H (2013) Plant apparency, an overlooked driver of associational resistance to insect herbivory. J Ecol 101:418–429. https://doi.org/10.1111/1365-2745.12055

Castagneyrol B, Jactel H, Vacher C, Brockerhoff EG, Koricheva J (2014a) Effects of plant phylogenetic diversity on herbivory depend on herbivore specialization. J Appl Ecol 51:134–141. https://doi.org/10.1111/1365-2664.12175

Castagneyrol B, Jactel H, Vacher C, Brockerhoff EG, Koricheva J, Cadotte M (2014b) Effects of plant phylogenetic diversity on herbivory depend on herbivore specialization. J Appl Ecol 51:134–141. https://doi.org/10.1111/1365-2664.12175

Chaplin-Kramer R, O'Rourke ME, Blitzer EJ, Kremen C (2011) A meta-analysis of crop pest and natural enemy response to landscape complexity. Ecol Lett 14:922–932

Cherif M, Loreau M (2013) Plant–herbivore–decomposer stoichiometric mismatches and nutrient cycling in ecosystems. Proc Royal Soc 280:20122453

Coley PD, Barone JA (1996) Herbivory and plant defenses in tropical forests. Annu Rev Ecol Syst 27:305–335. https://doi.org/10.1146/annurev.ecolsys.27.1.305

Deutsch CA, Tewksbury JJ, Tigchelaar M et al (2018) Increase in crop losses to insect pests in a warming climate. Science 361:916–919

Díaz S, Purvis A, Cornelissen JH et al (2013) Functional traits, the phylogeny of function, and ecosystem service vulnerability. Ecol Evol 3:2958–2975

Dinnage R, Cadotte MW, Haddad NM, Crutsinger GM, Tilman D (2012) Diversity of plant evolutionary lineages promotes arthropod diversity. Ecol Lett 15:1308–1317. https://doi.org/10.1111/j.1461-0248.2012.01854.x

Eisenhauer N, Schielzeth H, Barnes AD et al (2019) A multitrophic perspective on biodiversity–ecosystem functioning research. Academic Press, Adv. Ecol. Res

Endara M-J, Coley PD (2011) The resource availability hypothesis revisited: a meta-analysis. Funct Ecol 25:389–398. https://doi.org/10.1111/j.1365-2435.2010.01803.x

Engelhardt EK, Biber MF, Dolek M et al (2022) Consistent signals of a warming climate in occupancy changes of three insect taxa over 40 years in Central Europe. Global Change 28:3998–4012

Feeny P (1970) Seasonal changes in oak leaf tannins and nutrients as a cause of spring feeding by winter moth caterpillars. Ecology 51:565–581

Frost CJ, Hunter MD (2004) Insect canopy herbivory and frass deposition affect soil nutrient dynamics and export in oak mesocosms. Ecology 85:3335–3347

Futuyma DJ, Agrawal AA (2009) Macroevolution and the biological diversity of plants and herbivores. Proc Natl Acad Sci 106:18054–18061. https://doi.org/10.1073/pnas.0904106106

Getman-Pickering ZL, Soltis GJ, Shamash S, Gruner DS, Weiss MR, Lill JT (2023) Periodical cicadas disrupt trophic dynamics through community-level shifts in avian foraging. Science 382:320–324

Gilarranz LJ, Rayfield B, Liñán-Cembrano G, Bascompte J, Gonzalez A (2017) Effects of network modularity on the spread of perturbation impact in experimental metapopulations. Science 357:199–201

Giling DP, Ebeling A, Eisenhauer N et al (2019) Plant diversity alters the representation of motifs in food webs. Nat Commun 10:1226

Gómez JM, González-Megías A, Armas C, Narbona E, Navarro L, Perfectti F (2023) The role of phenotypic plasticity in shaping ecological networks. Ecol Lett 26:S47–S61

Gottschall F, Cesarz S, Auge H et al (2022) Spatiotemporal dynamics of abiotic and biotic properties explain biodiversity–ecosystem-functioning relationships. Ecol Monogr 92:e01490. https://doi.org/10.1002/ecm.1490

Guo P-F, Wang M-Q, Orr M et al (2021) Tree diversity promotes predatory wasps and parasitoids but not pollinator bees in a subtropical experimental forest. Basic Appl Ecol 53:134–142. https://doi.org/10.1016/j.baae.2021.03.007

Gurr GM, Wratten SD, Landis DA, You M (2017) Habitat management to suppress pest populations: progress and prospects. Annu Rev Entomol 62:91–109

Haase J, Castagneyrol B, Cornelissen JHC et al (2015) Contrasting effects of tree diversity on young tree growth and resistance to insect herbivores across three biodiversity experiments. Oikos 124:1674–1685. https://doi.org/10.1111/oik.02090

Haddad NM, Crutsinger GM, Gross K, Haarstad J, Knops JM, Tilman D (2009) Plant species loss decreases arthropod diversity and shifts trophic structure. Ecol Lett 12:1029–1039. https://doi.org/10.1111/j.1461-0248.2009.01356.x

Haddad NM, Crutsinger GM, Gross K, Haarstad J, Tilman D (2011) Plant diversity and the stability of foodwebs. Ecol Lett 14:42–46. https://doi.org/10.1111/j.1461-0248.2010.01548.x

Hagen SB, Jepsen JU, Ims RA, Yoccoz NG (2007) Shifting altitudinal distribution of outbreak zones of winter moth Operophtera brumata in sub-arctic birch forest: a response to recent climate warming? Ecography 30:299–307. https://doi.org/10.1111/j.2007.0906-7590.04981.x

Hance T, van Baaren J, Vernon P, Boivin G (2007) Impact of extreme temperatures on parasitoids in a climate change perspective. Annu Rev Entomol 52:107–126

Hättenschwiler S, Gasser P (2005) Soil animals alter plant litter diversity effects on decomposition. Proc Natl Acad Sci 102:1519–1524. https://doi.org/10.1073/pnas.0404977102

Hautier Y, Seabloom EW, Borer ET et al (2014) Eutrophication weakens stabilizing effects of diversity in natural grasslands. Nature 508:521–525

Haynes AG, Schütz M, Buchmann N, Page-Dumroese DS, Busse MD, Risch AC (2014) Linkages between grazing history and herbivore exclusion on decomposition rates in mineral soils of subalpine grasslands. Plant Soil 374:579–591

Huang Y, Chen Y, Castro-Izaguirre N et al (2018) Impacts of species richness on productivity in a large-scale subtropical forest experiment. Science 362:80–83. https://doi.org/10.1126/science.aat6405

Hunter MD (2001) Insect population dynamics meets ecosystem ecology: effects of herbivory on soil nutrient dynamics. Agric Forest Entomol 3

Hunter MD, Reynolds BC, Hall MC, Frost CJ (2012) Effects of herbivores on terrestrial ecosystem processes: the role of trait-mediated indirect effects. Trait-mediated indirect interactions: ecological and evolutionary perspectives. In: Trait-mediated indirect interactions: ecological and evolutionary perspectives, pp 339–370

Hwang BC, Giardina CP, Adu-Bredu S et al (2024) The impact of insect herbivory on biogeochemical cycling in broadleaved forests varies with temperature. Nat Commun 15:6011

Isbell F, Craven D, Connolly J et al (2015) Biodiversity increases the resistance of ecosystem productivity to climate extremes. Nature 526:574–577

Jactel H, Brockerhoff EG (2007) Tree diversity reduces herbivory by forest insects. Ecol Lett 10:835–848. https://doi.org/10.1111/j.1461-0248.2007.01073.x

Jactel H, Moreira X, Castagneyrol B (2021) Tree diversity and forest resistance to insect pests: patterns, mechanisms, and prospects. Annu Rev Entomol 66:277–296. https://doi.org/10.1146/annurev-ento-041720-075234

Jepsen J, Hagen S, Høgda K et al (2009) Monitoring the spatio-temporal dynamics of geometrid moth outbreaks in birch forest using MODIS-NDVI data. Remote Sens Environ 113:1939–1947

Johnson HE, Golden TS, Adams LG, Gustine DD, Lenart EA, Barboza PS (2021) Dynamic selection for forage quality and quantity in response to phenology and insects in an Arctic ungulate. Ecol Evol 11:11664–11688

Jorge LR, Novotny V, Segar ST et al (2017) Phylogenetic trophic specialization: a robust comparison of herbivorous guilds. Oecologia 185:551–559
Kagata H, Ohgushi T (2012) Non-additive effects of leaf litter and insect frass mixture on decomposition processes. Ecol Res 27:69–75. https://doi.org/10.1007/s11284-011-0868-6
Kessler A, Mueller MB, Kalske A, Chautá A (2023) Volatile-mediated plant–plant communication and higher-level ecological dynamics. Curr Biol 33:R519–R529
Kharouba HM, Ehrlén J, Gelman A et al (2018) Global shifts in the phenological synchrony of species interactions over recent decades. Proc Natl Acad Sci 115:5211–5216
Kigathi RN, Weisser WW, Reichelt M, Gershenzon J, Unsicker SB (2019) Plant volatile emission depends on the species composition of the neighboring plant community. BMC Plant Biol 19. https://doi.org/10.1186/s12870-018-1541-9
Knapp AK, Blair JM, Briggs JM et al (1999) The keystone role of bison in north American tallgrass prairie: bison increase habitat heterogeneity and alter a broad array of plant, community, and ecosystem processes. Bioscience 49:39–50
Knapp AK, Smith MD, Hobbie SE et al (2012) Past, present, and future roles of long-term experiments in the LTER network. Bioscience 62:377–389
König S, Krauss J, Keller A, Bofinger L, Steffan-Dewenter I (2022) Phylogenetic relatedness of food plants reveals highest insect herbivore specialization at intermediate temperatures along a broad climatic gradient. Global Change 28:4027–4040
Kristensen JÅ, Rousk J, Metcalfe DB (2020) Below-ground responses to insect herbivory in ecosystems with woody plant canopies: a meta-analysis. J Ecol 108:917–930. https://doi.org/10.1111/1365-2745.13319
Lemoine NP, Burkepile DE (2012) Temperature-induced mismatches between consumption and metabolism reduce consumer fitness. Ecology 93:2483–2489
Letourneau DK, Armbrecht I, Rivera BS et al (2011) Does plant diversity benefit agroecosystems? A synthetic review. Ecol Appl 21:9–21
Li Y, Schmid B, Schuldt A et al (2023) Multitrophic arthropod diversity mediates tree diversity effects on primary productivity. Nat. Ecol. Evol.:1–9
Li Y, Schuldt A, Ebeling A et al (2024) Plant diversity enhances ecosystem multifunctionality via multitrophic diversity. Nat. Ecol. Evol. 8:2037–2047
Liang M, Yang Q, Chase JM et al (2025) Unifying spatial scaling laws of biodiversity and ecosystem stability. Science 387:eadl2373
Ling S, Rizvi SAH, Xiong T et al (2022) Volatile signals from guava plants prime defense signaling and increase jasmonate-dependent herbivore resistance in neighboring citrus plants. Front Plant Sci 13:833562
lM-Arnold A, Grüning M, Simon J, Reinhardt A-B, Lamersdorf N, Thies C (2016) Forest defoliator pests alter carbon and nitrogen cycles. R Soc Open Sci 3:160361
McNaughton SJ (1984) Grazing lawns: animals in herds, plant form, and coevolution. Am Nat 124:863–886. https://doi.org/10.1086/284321
Medina-van Berkum P, Albracht C, Bröcher M et al (2025) Plant diversity shapes plant volatile emission differently at the species and community level. bioRxiv:2025.2004. 2030.651392
Metcalfe DB, Asner GP, Martin RE et al (2014) Herbivory makes major contributions to ecosystem carbon and nutrient cycling in tropical forests. Ecol Lett 17:324–332
Mohammed M (2024) Transient patterns and delayed herbivore response in plant-herbivore systems. Theor Ecol 17:235–245
Mooney KA, Pratt RT, Singer MS (2012) The tri-trophic interactions hypothesis: interactive effects of host plant quality, diet breadth and natural enemies on herbivores. PLoS One 7:e34403. https://doi.org/10.1371/journal.pone.0034403
Muiruri EW, Barantal S, Iason GR, Salminen JP, Perez-Fernandez E, Koricheva J (2019) Forest diversity effects on insect herbivores: do leaf traits matter? New Phytol 221:2250–2260. https://doi.org/10.1111/nph.15558
Neff F, Brändle M, Ambarlı D et al (2021) Changes in plant-herbivore network structure and robustness along land-use intensity gradients in grasslands and forests. Sci Adv 7:eabf3985. https://doi.org/10.1126/sciadv.abf3985

Novotny V, Basset Y, Miller SE et al (2002) Low host specificity of herbivorous insects in a tropical forest. Nature 416:841–844

Oerke E-C (2006) Crop losses to pests. J Agric Sci 144:31–43

Osterrieth M, Bosker T (2024) Impacts of terrestrial mammalian herbivores on vegetation change in the arctic. Polar Biol 47:1307–1321

Parmesan C, Hanley ME (2015) Plants and climate change: complexities and surprises. Ann Bot 116:849–864

Pearse IS, Hipp AL (2009) Phylogenetic and trait similarity to a native species predict herbivory on non-native oaks. Proc Natl Acad Sci 106:18097–18102

Pearse IS, Hipp AL (2012) Global patterns of leaf defenses in OAK species. Evolution 66:2272–2286. https://doi.org/10.1111/j.1558-5646.2012.01591.x

Pellissier L, Descombes P, Hagen O et al (2018) Growth-competition-herbivore resistance trade-offs and the responses of alpine plant communities to climate change. Funct Ecol 32:1693–1703. https://doi.org/10.1111/1365-2435.13075

Peltonen M, Liebhold AM, Bjørnstad ON, Williams DW (2002) Spatial synchrony in forest insect outbreaks: roles of regional stochasticity and dispersal. Ecology 83:3120–3129

Pilotto F, Kühn I, Adrian R et al (2020) Meta-analysis of multidecadal biodiversity trends in Europe. Nat Commun 11:3486

Pinho BX, Dattilo W, Leal IR (2018) Structural breakdown of specialized plant-herbivore interaction networks in tropical forest edges. Global Ecol Conserv 12:1–8

Popkin G (2021) Forest fight. Science 374:1184–1189. https://doi.org/10.1126/science.acx9733

Prather C, Strickland MS, Laws A, Branson D (2017) Herbivore species identity and composition affect soil enzymatic activity through altered plant composition in a coastal tallgrass prairie. Soil Biol Biochem 112:277–280

Pureswaran DS, Roques A, Battisti A (2018) Forest insects and climate change. Curr For Rep 4:35–50. https://doi.org/10.1007/s40725-018-0075-6

Rezende EL, Albert EM, Fortuna MA, Bascompte J (2009) Compartments in a marine food web associated with phylogeny, body mass, and habitat structure. Ecol Lett 12:779–788

Root RB (1973) Organization of a plant-arthropod association in simple and diverse habitats: the fauna of collards (Brassica oleracea). Ecol Monogr 43:95–124

Salazar D, Marquis RJ (2022) Testing the role of local plant chemical diversity on plant–herbivore interactions and plant species coexistence. Ecology 103:e3765

Schaefer HM, Ruxton GD (2011) Plant-animal communication. Oxford University Press, Oxford

Schaefer KVR, Clark KL, Skowronski N, Hamerlynck EP (2010) Impact of insect defoliation on forest carbon balance as assessed with a canopy assimilation model. Global Change 16:546–560. https://doi.org/10.1111/j.1365-2486.2009.02037.x

Scherber C, Eisenhauer N, Weisser WW et al (2010a) Bottom-up effects of plant diversity on multitrophic interactions in a biodiversity experiment. Nature 468:553–556. https://doi.org/10.1038/nature09492

Scherber C, Heimann J, Koeler G, Mitschunas N, Weisser WW (2010b) Functional identity versus species richness: herbivory resistance in plant communities. Oecologia 163:707–717. https://doi.org/10.1007/s00442-010-1625-1

Schmitz OJ (2008) Herbivory from individuals to ecosystems. Annu Rev Ecol Evol Syst 39:133–152

Schnabel F, Liu X, Kunz M et al (2021) Species richness stabilizes productivity via asynchrony and drought-tolerance diversity in a large-scale tree biodiversity experiment. Sci Adv 7:eabk1643

Schnabel F, Beugnon R, Yang B et al (2025) Tree diversity increases forest temperature buffering via enhancing canopy density and structural diversity. Ecol Lett 28:e70096

Schowalter TD (2012) Insect herbivore effects on forest ecosystem services. J Sustain For 31:518–536. https://doi.org/10.1080/10549811.2011.636225

Schuldt A, Baruffol M, Bohnke M et al (2010) Tree diversity promotes insect herbivory in subtropical forests of south-East China. J Ecol 98:917–926. https://doi.org/10.1111/j.1365-2745.2010.01659.x

Schuldt A, Bruelheide H, Durka W et al (2012) Plant traits affecting herbivory on tree recruits in highly diverse subtropical forests. Ecol Lett 15:732–739. https://doi.org/10.1111/j.1461-0248.2012.01792.x
Schuldt A, Assmann T, Brezzi M et al (2018) Biodiversity across trophic levels drives multifunctionality in highly diverse forests. Nat Commun 9:2989. https://doi.org/10.1038/s41467-018-05421-z
Schuldt A, Ebeling A, Kunz M et al (2019) Multiple plant diversity components drive consumer communities across ecosystems. Nat Commun 10:1460. https://doi.org/10.1038/s41467-019-09448-8
Seidl R, Thom D, Kautz M et al (2017) Forest disturbances under climate change. Nat Clim Chang 7:395. https://doi.org/10.1038/nclimate3303
Shen Z-Q, Zhu X-H, Wang M-Q et al (2025) Effects of neighborhood diversity on herbivory in early-stage forests in southern China. Plant Diver. https://doi.org/10.1016/j.pld.2025.09.008
Siemann E, Tilman D, Haarstad J, Ritchie M (1998) Experimental tests of the dependence of arthropod diversity on plant diversity. Am Nat 152:738–750. https://doi.org/10.1086/286204
Singer MC, Parmesan C (2010) Phenological asynchrony between herbivorous insects and their hosts: signal of climate change or pre-existing adaptive strategy? Philos Trans R Soc B Biol Sci 365:3161–3176
Soliveres S, van der Plas F, Manning P et al (2016) Biodiversity at multiple trophic levels is needed for ecosystem multifunctionality. Nature 536:456–459. https://doi.org/10.1038/nature19092
Srednick G, Swearer SE (2024) Understanding diversity–synchrony–stability relationships in multitrophic communities. Nat Ecol Evol:1–11
Staab M, Blüthgen N, Klein A-M (2015) Tree diversity alters the structure of a tri-trophic network in a biodiversity experiment. Oikos 124:827–834. https://doi.org/10.1111/oik.01723
Staab M, Bruelheide H, Durka W et al (2016) Tree phylogenetic diversity promotes host-parasitoid interactions. Proc. R. Soc. 283. https://doi.org/10.1098/rspb.2016.0275
Staab M, Liu X, Assmann T et al (2020) Tree phylogenetic diversity structures multitrophic communities. Funct Ecol 35:521–534. https://doi.org/10.1111/1365-2435.13722
Tanentzap AJ, Coomes DA (2012) Carbon storage in terrestrial ecosystems: do browsing and grazing herbivores matter? Biol Rev 87:72–94
Tang T, Zhang N, Bongers FJ et al (2022) Tree species and genetic diversity increase productivity via functional diversity and trophic feedbacks. elife 11. https://doi.org/10.7554/eLife.78703
Tenow O, Bylund H, Nilssen A, Karlsson PS (2005) Long-term influence of herbivores on northern birch forests. In: Plant ecology, herbivory, and human impact in Nordic mountain birch forests. Springer, pp 165–181
Terhorst CP, Zee PC, Heath KD et al (2018) Evolution in a community context: trait responses to multiple species interactions. Am Nat 191:368–380
Thebault E, Fontaine C (2010) Stability of ecological communities and the architecture of mutualistic and trophic networks. Science 329:853–856. https://doi.org/10.1126/science.1188321
Thébault E, Loreau M (2006) The relationship between biodiversity and ecosystem functioning in food webs. Ecol Res 21:17–25
Tilman D, Downing JA (1994) Biodiversity and stability in grasslands. Nature 367:363–365. https://doi.org/10.1038/367363a0
Tilman D, Reich PB, Knops JM (2006) Biodiversity and ecosystem stability in a decade-long grassland experiment. Nature 441:629–632
Tscharntke T, Klein AM, Kruess A, Steffan-Dewenter I, Thies C (2005) Landscape perspectives on agricultural intensification and biodiversity–ecosystem service management. Ecol Lett 8:857–874. https://doi.org/10.1111/j.1461-0248.2005.00782.x
Turcotte MM, Davies TJ, Thomsen CJM, Johnson MTJ (2014) Macroecological and macroevolutionary patterns of leaf herbivory across vascular plants. Proc R Soc B Biol Sci 281. https://doi.org/10.1098/rspb.2014.0555
Turlings TC, Erb M (2018) Tritrophic interactions mediated by herbivore-induced plant volatiles: mechanisms, ecological relevance, and application potential. Annu Rev Entomol 63:433–452

Tylianakis JM, Morris RJ (2017) Ecological networks across environmental gradients. Annu Rev Ecol Evol Syst 48:25–48

Tylianakis JM, Didham RK, Bascompte J, Wardle DA (2008) Global change and species interactions in terrestrial ecosystems. Ecol Lett 11:1351–1363. https://doi.org/10.1111/j.1461-0248.2008.01250.x

van Asch M, Visser ME (2007) Phenology of forest caterpillars and their host trees: the importance of synchrony. Annu Rev Entomol 52:37–55

van der Plas F (2019) Biodiversity and ecosystem functioning in naturally assembled communities. Biol Rev Camb Philos Soc 94:1220–1245. https://doi.org/10.1111/brv.12499

van Klink R, Lepš J, Vermeulen R, de Bello F (2019) Functional differences stabilize beetle communities by weakening interspecific temporal synchrony. Ecology 100:e02748. https://doi.org/10.1002/ecy.2748

Vehvilainen H, Koricheva J, Ruohomaki K (2007) Tree species diversity influences herbivore abundance and damage: meta-analysis of long-term forest experiments. Oecologia 152:287–298. https://doi.org/10.1007/s00442-007-0673-7

Visakorpi K, Gripenberg S, Malhi Y, Riutta T (2024) Low-intensity insect herbivory could have large effects on ecosystem productivity through reduced canopy photosynthesis. Web Ecol 24:97–113

Voglar GE, Mrak T, Krizman M, Jagodic A, Trdan S, Laznik Z (2019) Effect of contaminated soil on multitrophic interactions in a terrestrial system. Plant Soil 435:337–351. https://doi.org/10.1007/s11104-018-03903-z

Volf M, Segar ST, Miller SE et al (2018) Community structure of insect herbivores is driven by conservatism, escalation and divergence of defensive traits in Ficus. Ecol Lett 21:83–92. https://doi.org/10.1111/ele.12875

Wang MQ, Li Y, Chesters D et al (2019) Multiple components of plant diversity loss determine herbivore phylogenetic diversity in a subtropical forest experiment. J Ecol 107:2697–2712. https://doi.org/10.1111/1365-2745.13273

Wang MQ, Li Y, Chesters D et al (2020) Host functional and phylogenetic composition rather than host diversity structure plant-herbivore networks. Mol Ecol 29:2747–2762. https://doi.org/10.1111/mec.15518

Wang M-Q, Yan C, Luo A et al (2022) Phylogenetic relatedness, functional traits, and spatial scale determine herbivore co-occurrence in a subtropical forest. Ecol Monogr 92:e01492. https://doi.org/10.1002/ecm.1492

Wang M-Q, Wen Z, Ke J et al (2023) Tree communities and functional traits determine herbivore compositional turnover. Oecologia 203:205–218. https://doi.org/10.1007/s00442-023-05463-1

Wang M-Q, Albert G, Chesters D et al (2025a) Tree diversity, tree growth, and microclimate independently structure Lepidoptera herbivore community stability. Ecol Monogr 95:e70026. https://doi.org/10.1002/ecm.70026

Wang M-Q, Guo S-K, Guo P-F et al (2025b) Multidimensionality of tree communities structure host-parasitoid networks and their phylogenetic composition. elife 13:RP100202. https://doi.org/10.7554/eLife.100202

Wang M-Q, Zhang K, Albert G et al (2025c) Cascading effects of tree diversity loss on herbivore-parasitoid interactions. Biol Conserv 311:111418

Wang M-Q, Albert G, Seifert CL et al (2026) Asynchrony and functional diversity couple herbivore community dynamics to host plant diversity. Nat Commun 17

Wardle DA, Yeates GW, Williamson WM, Bonner KI, Barker GM (2004) Linking aboveground and belowground communities: the indirect influence of aphid species identity and diversity on a three trophic level soil food web. Oikos 107:283–294. https://doi.org/10.1111/j.0030-1299.2004.13523.x

Weiblen GD, Webb CO, Novotny V, Basset Y, Miller SE (2006) Phylogenetic dispersion of host use in a tropical insect herbivore community. Ecology 87:S62–S75

Welti EA, Kaspari M (2021) Sodium addition increases leaf herbivory and fungal damage across four grasslands. Funct Ecol 35:1212–1221

Wetzel WC, Inouye BD, Hahn PG, Whitehead SR, Underwood N (2023) Variability in plant–herbivore interactions. Annu Rev Ecol Evol Syst 54:451–474
White TC (2012) The inadequate environment: nitrogen and the abundance of animals. Springer Science & Business Media
Wright IJ, Reich PB, Westoby M et al (2004) The worldwide leaf economics spectrum. Nature 428:821–827
Zhang J, Bruelheide H, Chen X et al (2017) Tree diversity promotes generalist herbivore community patterns in a young subtropical forest experiment. Oecologia 183:455–467. https://doi.org/10.1007/s00442-016-3769-0
Zu P, Boege K, Del-Val E et al (2020) Information arms race explains plant-herbivore chemical communication in ecological communities. Science 368:1377–1381

Direct and Indirect Effects of Global Change on Herbivore Communities

5

Anne Kempel and Eric Allan

Abstract

Global change is reshaping herbivore communities, with important implications for ecosystem functioning and the regulation of plant diversity. Global change drivers can influence herbivores both directly, by altering their physiology, survival, or population dynamics, and indirectly. Indirect effects can arise through multiple pathways, but changes in plant communities are of major importance as plants constitute the food resources and habitat for herbivores. In this chapter, we synthesise ecological theory and empirical evidence on how major global change drivers—including climate warming, altered precipitation, elevated atmospheric CO_2, land-use intensification, and biological invasions—affect herbivore communities through these two pathways. We examine plant diversity as a key mediator of indirect effects while also highlighting the importance of additional plant community dimensions that shape herbivore communities, including plant functional trait composition, biomass production, and tissue quality. We show that some global change drivers, e.g. warming and mowing or grazing, can exert strong direct effects on herbivore physiology and survival, whereas drivers such as nitrogen enrichment and elevated CO_2 primarily influence herbivores indi-

A. Kempel (✉)
WSL Institute for Snow and Avalanche Research SLF, Davos, Switzerland

Climate Change, Extremes and Natural Hazards in Alpine Regions Research Centre CERC, Davos, Switzerland
e-mail: anne.kempel@slf.ch

E. Allan
Institute of Plant Sciences and Oeschger Centre for Climate Change Research, University of Bern, Bern, Switzerland

Centre for Development and Environment, University of Bern, Bern, Switzerland

X. Moreira et al. (eds.), *Effects of Plant Diversity on Plant-Herbivore Interactions*, Ecological Studies 255, https://doi.org/10.1007/978-3-032-26461-9_5

rectly through effects of changes in plant diversity, productivity, nutritional quality, and functional composition. Because these plant community dimensions are tightly linked and highly sensitive to environmental change, indirect effects can either amplify or dampen direct effects of global change on herbivores. We also consider reciprocal effects of herbivores on plant communities, highlighting how herbivores can modify plant responses to global change. We discuss methodological approaches for disentangling direct and indirect effects and outline implications for ecosystem management under ongoing environmental change. By integrating direct and indirect, plant-mediated pathways, this chapter provides a holistic framework for predicting herbivore responses to global change across ecosystems and management contexts.

5.1 Understanding Global Change Effects on Herbivores: Direct and Indirect Effects

As the pace of environmental change accelerates, predicting how biological communities and ecosystem functioning will respond is a central challenge in ecology (Gilman et al. 2010; Parmesan 2006). Within this context, understanding future changes in interactions between plants and their consumers, such as herbivores, is particularly important, because shifts in herbivory can profoundly influence plant biodiversity, productivity, and even food security (Bebber et al. 2013; Deutsch et al. 2018). Global change drivers, including climate warming, rising atmospheric CO_2, altered precipitation regimes, nutrient enrichment, and land-use intensification, can affect herbivores directly, by altering their physiology, behaviour, survival, or population dynamics, or indirectly, for instance, by altering the plants that herbivores feed on or the predators that attack herbivores (Suttle et al. 2007; Tylianakis et al. 2008, Fig. 5.1). For example, warming, changes in precipitation, and alterations in management practices often have immediate, direct impacts on herbivores (Cornelissen 2011; Hamann et al. 2021; Humbert et al. 2010), whereas nitrogen enrichment and elevated CO_2 primarily influence herbivore communities indirectly by altering plant diversity and other community attributes such as functional trait composition, nutritional quality, and biomass production (Bobbink et al. 2010; Kaspari and Welti 2024). Other indirect pathways, such as effects mediated through above- or belowground microbial communities, may also play important roles but are not considered in detail here.

Understanding indirect effects mediated via plant community attributes is particularly critical, because herbivore communities are strongly structured by plants, and plant communities themselves are highly sensitive to environmental change (Franklin et al. 2016). Plant biodiversity loss is one key pathway through which global change indirectly affects herbivores, and a large body of literature has looked at how plant diversity impacts herbivores (the main focus of this volume). However, changes in other plant community attributes—such as shifts in functional composition, plant quality, and productivity—can also strongly influence herbivores. Many of these changes in community attributes may be correlated (Allan et al. 2015), and

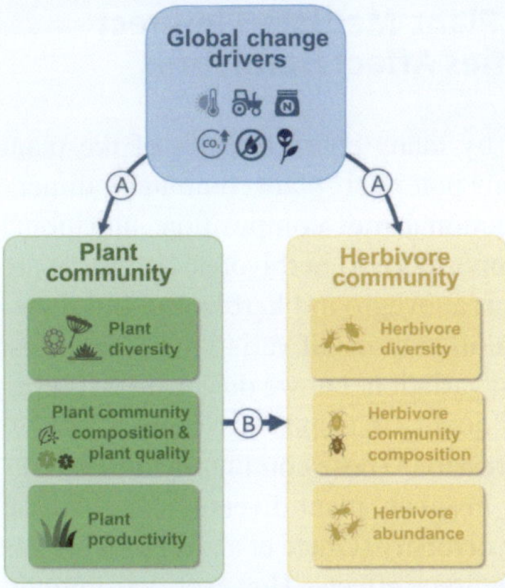

Driver	(A) Direct effects on herbivores	(B) Indirect effects via plants
Warming temperature	***Positive***: increases metabolic rates, growth and reproduction	***Positive***: increases in plant productivity can enhance food availability, thereby benefiting herbivores through bottom-up processes
Drought	***Varying***: effects depending on herbivore species	***Positive short-term effects***: Drought can reduce plant defenses, potentially increasing herbivore feeding; ***negative long-term effects***: Plant community shifts towards slow-growing, well-defended species, which may ultimately decrease herbivore abundance
Elevated CO2	No direct effects	***Negative***: lower plant nutritional quality (via nutrient dilution and altered stoichiometry), which may reduce herbivore growth and abundance
Mowing	***Negative***: mechanical damage and killing of herbivores	***Positive at low intensity***: maintains plant diversity and thus supports herbivore diversity; ***negative at high intensity***: removes food resources and reduces plant diversity, decreasing herbivore diversity and abundance.
Grazing	***Negative***: physical damage and trampling	***Positive at low intensity***: Enhances heterogeneity and plant diversity, supporting herbivores; ***negative at high intensity***: Reduces food and habitat (via biomass removal and soil compaction, lowering herbivore abundance and diversity
Nitrogen fertilisation	No direct effects	***Negative***: Favours fast-growing species and reduces plant diversity, lowering herbivore diversity; ***potentially positive***: increases herbivore abundance through increasing plant nutrient contents and shifts towards fast-growing, palatable plant species
Pesticide Use	***Negative***: toxic effects on herbivores	
Invasive species	***Negative***: poor host quality, especially for specialists herbivores	***Negative:*** displace native plants, reducing high-quality food availability

Fig. 5.1 Global change drivers such as increases in temperature, changes in precipitation such as drought, elevated CO_2, land-use intensification through nitrogen enrichment, grazing, mowing or pesticide use, and invasive plant species can *directly* affect different aspects of plant and herbivore communities (A arrows). In addition, changes in the plant community can *indirectly* affect herbivores, e.g. changes in plant diversity, plant community composition and plant quality, and plant productivity may cascade up and affect different aspects of the herbivory community, such as herbivore diversity, herbivore community composition, i.e. whether herbivores are dietary generalists or specialists (G vs S), and herbivore abundance (B arrow)

it is therefore critical to separate their different effects to understand their relative importance and to understand the importance of plant diversity change in impacting herbivore communities. Plant-mediated indirect effects can also amplify or dampen the direct effects of global change on herbivores, often leading to complex and sometimes unexpected outcomes (Paseka et al. 2020; Post and Pedersen 2008). This makes it critical to separate the direct and indirect effects, because overall global change effects may, for example, appear weak if direct and indirect effects strongly interact to reduce each other's effects.

This chapter synthesises ecological theory and empirical evidence on how global change drivers affect herbivore communities and their impact on plants (i.e. via herbivory) through both direct and indirect pathways (Fig. 5.1). We focus on indirect effects driven via changes in plant diversity but also consider other plant attributes such as functional composition, quality, and productivity. In addition, we address the reciprocal effects of herbivores on plant communities, highlighting how herbivores can modify plant responses to global change and generate feedbacks between plants and consumers. By integrating these different perspectives, we aim to provide a holistic view of how environmental change reshapes species interactions and influences broader ecosystem processes. Before addressing global change drivers, we start by outlining the fundamental ways in which plant communities structure herbivore communities, providing the ecological basis for understanding indirect, plant-mediated responses to environmental change.

5.2 Understanding the Basics of Plant-Mediated Indirect Effects: How Plant Communities Affect Herbivores

Herbivore communities are strongly shaped by many characteristics of the plant communities they feed on. While there are many potential "plant community dimensions" that could be explored, diversity, functional trait composition, nutritional quality, and biomass production are of key importance to herbivores, and these are the dimensions we will focus on. We focus on aboveground herbivores and therefore mostly consider functional trait composition in terms of variation in leaf traits. Nutritional quality is clearly linked to leaf trait variation, but we do consider it separately, because plants may rapidly change leaf quality in response to nutrient or CO_2 enrichment, before they change their trait expression. These community dimensions can also be causally linked to each other; for instance, plant diversity can increase biomass, while high biomass can reduce plant diversity (Grace et al. 2016). This has the potential to lead to complex chains of indirect effects. However, we initially treat the attributes as independent dimensions to explore their separate effects before considering how global change may alter several community dimensions at the same time. All of these plant community characteristics are themselves highly sensitive to environmental change and therefore represent key indirect pathways for impact on herbivores. In this section, we review ecological theory and empirical evidence on how these components of plant communities influence different properties of herbivore communities (i.e. herbivore diversity, abundance, and community composition; Table 5.1). Exploring the links between these plant and herbivore community dimensions provides the foundation for understanding the indirect effects of global change on herbivore dynamics.

5.2.1 The Effect of Plant Diversity on Herbivores

Plant diversity is generally thought to lead to high herbivore *diversity*, i.e. diversity begets diversity. Most herbivore species are specialist, i.e. feeding on a single plant family or species (Forister et al. 2015), and it is clear that a large portion of host plant species can support a greater diversity of specialist herbivores feeding on them (Scherber et al. 2010). Such a positive relationship between plant diversity and invertebrate diversity has been shown at local (Ampoorter et al. 2020; Brändle et al. 2001; Haddad et al. 2000; Murdoch et al. 1972; Scherber et al. 2010) and global scales (Coley and Kursar 2014).

The *abundance* and *composition* of herbivore communities can also be affected by plant diversity. Diverse plant communities generally have a more complex vegetation structure, which offers more microhabitats and refuges for herbivores, thereby enhancing abundance and facilitating coexistence (Lawton 1983; Southwood et al. 1979). Plant diversity may also affect herbivore abundance, because diverse plant communities provide a greater range of resources for herbivores. This high-resource diversity is likely to support a high diversity of specialist herbivores; however, the abundance of individual specialist herbivores is likely to remain low,

Table 5.1 Overview of hypotheses on how different aspects of plant communities, namely, plant diversity, plant biomass, plant functional composition (fast vs. slow-growing plant species, plant height) affect different aspects of herbivore communities such as abundance, diversity, and community composition (whether communities consist of rather generalist or specialist herbivores), and how different aspects of the herbivore community affect plants

Hypothesis and References	Predicted pattern
Effect of plant communities on herbivore abundance	
Bottom-up control of plant consumers Hunter & Price 1992; Haddad et al. 2001	A higher plant productivity increases herbivore abundance.
Top-down control of plant herbivores Hairston et al. 1960	If herbivores are regulated by their predators, then changes in plant productivity will not affect herbivore abundance.
Exploitation ecosystem hypothesis Fretwell 1987, Oksanen & Oksanen 2000	Changes in plant productivity alter the strength of top-down versus bottom-up regulation: Top-down control of predators on herbivores increases with plant productivity.
Resource concentration hypothesis Root 1973	At lower plant diversity, specialist herbivores increase as they can easily find and locate their host. In turn, a high plant diversity reduces the abundance of specialist herbivores.
Effects of diet mixing, resource diversity, consistent resource supply Price et al. 1980, Haddad et al. 2001, Pfisterer et al. 2003	Plant diversity can also increase herbivore abundance, and in particular the abundance of generalist herbivores as they might benefit from resource diversity and more consistent resource supply. Generalist herbivores can also more easily spill over to other hosts in diverse plant communities.
Plant architecture Lawton 1983; Southwood et al. 1979	Invertebrate herbivores benefit from a more complex vegetation structure in more diverse plant communities
Enemy hypothesis Root 1973; Barnes et al. 2020	Not only herbivores benefit from a higher structural complexity in diverste communities, but also their predators, and they reduce herbivore abundance in diverse plant communities.
Associational susceptibility Barbosa et al. 2009; Halliday et al. 2017	The presence of unpalatable plant species in a community decreases herbivore attack also on other species
Growth-defence trade-off hypothesis Coley et al. 1985; Cappelli et al. 2020	Plant communities consisting of more fast-growing but less defended plant species are expected to support higher herbivore abundances
Apparency hypothesis, Feeny 1976	Tall plants are more apparent, and might therefore attract more herbivores
Effects of plant communities on herbivore diversity	
More individuals hypothesis Srivastava & Lawton, 1998	A higher plant productivity promotes a higher herbivore abundance. A higher total number of herbivore individuals in a community allows for a greater number of species with viable populations.
Diversity promotes diversity e.g. Murdoch et al. 1972; Rottstock et al. 2014	Many herbivores are host specialists. A higher plant diversity therefore promotes a higher herbivore diversity.
Effects of plant communities on herbivore community composition	
Resource concentration hypothesis Root 1973	Specialist herbivores are expected to be most abundant in plant species poor communities, while generalist herbivores may increase at high plant diversity
Fast-growing vs slow growing plant communities (growth-defense trade-off) Züst & Agrawal 2017; Ali & Agrawal 2012	Generalist herbivores are suggested to respond more strongly to increases in plant quality. Hence, fast growing plant communities may support a higher abundance of generalist herbivores.
Effects of herbivores on plant productivity	
Top-down control Paine et al. 1966	A high herbivore abundance is expected to increase herbivore impact on plant productivity.
Biodiversity-ecosytem functioning theory Deraison et al. 2015; Mordecai, 2011, Ritchie &Olff 1999	A more diverse herbivore community might use plant resource more completely and will have a stronger negative impact on plant productivity.
Diversity suppresses virulence Johnson et al. 2013	A more diverse herbivore community could also decrease herbivore impact on plant productivity if herbivores compete with each other and a high diversity prevents the dominance of particularly virulent species.
Metabolic theory, Brown et al. 2014	Herbivore communities consisting of many small insects have a higher energy demand than herbivore communities consisting of few large insects (given their total biomas is the same).
Effects of herbivores on plant diversity	
Herbivores promote coexistence by affecting niche and fitness differences between plants E.g. Chesson 2000; Carson & Root 2000, Levine & HilleRisLambers 2009; Bagchi et al. 2014; Coley et al. 1985; Kempel et al. 2011, Cappelli et al. 2020	Herbivores can promote plant diversity if they more strongly attack the most abundant species in a community. They can stabilize coexistence if their attack rate increases with increasing host frequency. Herbivores can also promote coexistence by equalizing the competitive ability between plant species, e.g. when there is a trade off between growth and defense.
Herbivores can destabilize plant coexistence Allan & Crawley 2011	If herbivores more strongly attack the least abundant plant species they increase fitness differences among plant species which can decrease diversity.
Additive vs compensatory effects of herbivores Ritchie & Olff 1999	Depending on whether different herbivore species attack the same plant species (additive effects) or different plant species (compensatory effects), herbivore diversity can either reduce or enhance plant diveristy, respectively.
Effects of herbivores on plant composition	
Herbivores cause shifts in plant composition Crawley 1997	Plant herbivores, in particular generalists, may preferentially attack fast growing, less defended plant species rather than slower growing, more defended ones. This shifts plant communities towards slower growing, more defended species.

Effects driven by:

Plant productivity	Herbivore abundance
Plant diversity	Herbivore diversity
Plant community composition	Herbivore community composition (generalists, specialists)
Plant size	Herbivore size

because each individual plant species occurs only at low density and therefore cannot support a large population of its specialist herbivores (Root 1973). In contrast, generalist herbivores may profit from a diverse plant community because high-resource diversity offers greater opportunities for diet mixing, toxin dilution, and a stable food supply throughout the season (Bernays et al. 1994; Pfisterer et al. 2003; Price 1991). This is therefore likely to lead to a high abundance of generalist herbivores, and an increase in herbivore abundance in diverse plant communities also leads to an increase in generalist herbivore diversity (Borer et al. 2012). At the same time, diverse plant community may also provide more niches for predators, and this increased predator abundance may reduce overall herbivore abundance by increasing top-down control, and this is often referred to as the "Enemies Hypothesis" (Barnes et al. 2020; Root 1973). Several studies found that herbivore abundance increases with plant diversity (Ebeling et al. 2014; Haddad et al. 2001; Jactel and Brockerhoff 2007; Moreira et al. 2016; Scherber et al. 2010); however, others found opposing patterns (e.g. Koricheva et al. 2000). Hence, while a loss of plant diversity is likely to result in a loss of herbivore diversity, the effect of plant diversity on herbivore abundance is likely to depend on several factors, such as their host specificity and predator pressure.

5.2.2 The Effect of Plant Biomass Production on Herbivores

It is expected that changes in plant biomass production would affect the herbivores feeding on the plants. However, the extent and direction of effects depend on how herbivores are regulated, and there is a large body of literature predicting whether changes in plant biomass will impact herbivores. Herbivores will only respond to changes in plant biomass if they are limited by food availability, a scenario known as bottom-up regulation, in which higher plant productivity supports higher herbivore abundance (Hairston et al. 1960; Hunter and Price 1992). However, if herbivore populations are primarily controlled by predators (top-down regulation), then increases in plant biomass may not necessarily lead to higher herbivore abundance, or they may only weakly affect herbivore abundance. This idea was published as the "green world hypothesis" (Hairston et al. 1960), later on refined in the "exploitation ecosystem hypothesis", which predicts that top-down control strengthens with increasing resource availability: In low-productivity systems, herbivores are mainly limited by food, whereas in highly productive systems, abundant resources support larger predator populations, which can suppress herbivore numbers (Fretwell 1987; Oksanen and Oksanen 2000). Empirical tests of the exploitation ecosystem hypothesis have shown that its predictions do hold for vertebrate herbivores in some cases (Aunapuu et al. 2008; Oksanen et al. 2020). However, results for invertebrates are more mixed (Kempel et al. 2023; Schädler et al. 2003; Welti et al. 2020), suggesting that variation in top-down vs. bottom-up control is more challenging to predict for insects. Early studies tended to assume that insect herbivores must be largely regulated by their predators (Crawley 1983; Olff and Ritchie 1998); however, more recent studies show bottom-up effects and demonstrate that increases in plant biomass do often increase insect abundance in many cases (e.g. Chase et al. 2000; Kempel et al. 2023; Lind et al. 2017). It therefore seems likely that increases in plant productivity would often lead to higher herbivore abundance, although it still remains challenging to predict when predation might limit changes in herbivore abundance.

Changes in plant biomass may also shape herbivore *diversity*, for example, via increases in herbivore abundance. When herbivore numbers rise, the pool of individuals becomes large enough to sustain more species with viable populations. This idea is captured by the "more individuals hypothesis" (Srivastava and Lawton 1998), which suggests that increased plant growth does not only increase herbivore numbers but also the variety of species present. In addition, high plant biomass is also likely to increase structural heterogeneity, which may in turn lead to a higher abundance of herbivores (Lawton 1983) and therefore also higher diversity.

5.2.3 The Effects of Plant Quality and Community Composition on Herbivores

The quality of individual plants and the composition of plant communities also strongly shape herbivore communities. Insect herbivores typically perform better when they are able to consume a diet with higher resource quality, and this typically leads to a higher abundance of herbivores in a plant community with high-resource levels. Due to stoichiometric constraints, herbivores are typically limited by nitrogen or phosphorus content in plants. When plants have higher nutritional value, e.g. higher levels of nitrogen or phosphorus, herbivores grow faster and ultimately herbivore abundance increases (Mattson 1980; Throop and Lerdau 2004; White 1993). In addition to these macronutrients, resources like sodium may also be very important for insect herbivores (Welti et al. 2019). Plants growing in nutrient-rich soils generally have higher nutrient concentrations and would therefore be expected to support a higher load of herbivores, i.e. more herbivores per unit of plant biomass. In addition, overall nutrient levels also vary substantially among species due to differences in ecological strategies. Aboveground, the distinction between fast-growing acquisitive plant species, with high leaf turnover and photosynthetic rates, and slow-growing conservative species, is a key dimension of strategy variation (Díaz et al. 2016; Wright et al. 2004). Fast-growing acquisitive plant species are good at competing for light and typically have high leaf nitrogen values (Díaz et al. 2016). These effects scale up to the community level, so that plant communities dominated by fast-growing, palatable species, support a higher number of herbivores (Coley et al. 1985).

As well as being nutritious, fast-growing plants may also invest less in defence. Long-standing theory predicts that in high-resource environments, plants should invest most in growth and little in defence, while the opposite should be true in low-resource environments (growth-defence trade-off, Coley et al. 1985). Mechanistic evidence for this hypothesis was recently provided by a paper showing that fast growing plants (with a high specific leaf area) had fewer genes involved in immune defence (Giolai and Laine 2024), supporting the assumption that fast-growing plant communities might harbour more herbivores. However, in some high-resource environments, plants may invest heavily in nitrogen-based defensive compounds, and this can reduce the benefits of high nutritional quality, and hence the number of herbivores (Stamp 2003). Further, while plants with high levels of chemical defence may deter generalist herbivores, specialists may be able to evolve to overcome these chemical defences and might be attracted to plant communities dominated by slow-growing plants. Some herbivores even sequester these compounds in their own tissues, turning plant defences into protection against their own predators (Petschenka and Agrawal 2016). Slower-growing plants might therefore support a large number of specialist herbivores, which would suggest that plant functional composition, in terms of growth strategy, could link to herbivore community composition in terms of degree of specialisation.

5.3 Direct and Indirect Effects of Global Change Drivers on Herbivores

Many global change drivers will have rapid direct effects on herbivore communities. Direct impacts on herbivore metabolism or survival arising from changes in temperature or physical disturbance can have immediate effects. At the same time, global change also reshapes plant diversity, productivity, nutritional quality, and community composition, which in turn alter herbivore communities. Because these pathways often interact, understanding both is crucial for predicting future herbivore dynamics and their ecological consequences. In the following sections, we synthesise evidence from a wide range of studies to show how major global change drivers—including warming, altered precipitation patterns, elevated CO_2, land-use intensification through mowing, grazing, fertilisation, and pesticide use, and invasive plant species —affect herbivores through these intertwined mechanisms (Fig. 5.1).

5.3.1 Rising Temperature and Changes in Precipitation

Rising temperatures and shifting precipitation patterns are expected to reshape herbivore populations in profound ways (Hamann et al. 2021). Global climate models project a temperature increase of 2–4 °C above preindustrial levels by the end of this century (IPCC 2023), and such warming directly influences the metabolic rates of herbivorous ectotherms (Brown et al. 2004; Tylianakis et al. 2008). When temperatures remain within physiological tolerance limits, faster metabolism translates into accelerated development and reproduction (Coakley et al. 1999; Lemoine et al. 2013, 2014; Roy et al. 2004). As a result, warming is expected to lead to higher population growth rates, more generations per season, increased overwintering survival, and ultimately increased invertebrate herbivore abundance yearlong (Bale et al. 2002; Cornelissen 2011). Precipitation changes and extreme events, however, complicate the picture. Changes in soil and air humidity are likely to have different effects depending on the herbivore group. Moisture-dependent groups like molluscs may decline or reduce their feeding activity during drought (Mensink and Henry 2011; Thakur et al. 2018), while other taxa such as small insects or some lepidopteran larvae suffer from high levels of precipitation (Chen et al. 2019; Kobori and Amano 2003). Warming and drought are also likely to be associated in many cases. Extreme weather events such as storms may directly influence herbivores, facilitating dispersal for some species while reducing populations of others (Cornelissen 2011; Rosenzweig et al. 2001). Overall, direct effects of warming are expected to be largely positive for herbivores, whereas precipitation effects are more context-dependent and might depend on the herbivore species or guild.

Climate drivers also have strong indirect effects. Warming generally stimulates plant growth when water is not limiting (Dobson and Zarnetske 2025), potentially supporting larger herbivore populations through bottom-up effects (Crawley 1983). In contrast, changes in soil moisture can alter plant quality in the short term and shift community composition over longer timescales. Periodic drought often reduces plant growth, and therefore the ability to compensate for herbivore damage, or it impairs herbivore defences, making it easier for herbivores to attack plants; both processes can lead to herbivore outbreaks which strongly impact plant communities (Lantschner et al. 2019; Mattson and Haack 1987). Prolonged drought may favour slow-growing, stress-tolerant plants (Coley et al. 1985; Lind et al. 2013), though in some cases, fast-growing, drought-avoiding species can instead dominate (Yan et al. 2025). These contrasting plant strategies affect herbivore communities differently: Specialist herbivores may thrive on slow-growing, chemically defended hosts, while generalists benefit from fast-growing, poorly defended plants. Furthermore, both warming and altered precipitation can also shift predator populations (Franco et al. 2019; Laws 2017). For instance, Arctic wolf spiders (*Pardosa glacialis*) have been observed producing additional clutches in years with earlier snowmelt, increasing their abundance and reshaping tundra food webs (Høye et al. 2020). Hence, although warming (if it occurs without an increase in drought) may increase herbivore populations through direct physiological effects and plant-mediated bottom-up pathways, these benefits may be indirectly offset if warming simultaneously enhances predator populations. In contrast, the consequences of drought or altered precipitation levels are more variable and depend on timescale, plant community responses, and the particular herbivore groups involved.

5.3.2 Elevated CO_2

Unlike warming, elevated CO_2 has little to no direct physiological impact on herbivores; instead, its consequences are mediated almost entirely through plants. Elevated CO_2 generally stimulates plant growth (Ainsworth and Long 2004), but this apparent boon for herbivores comes with hidden costs: higher tissue C:N ratios, and greater investment in carbon-based secondary and structural compounds, dilute nutrients and reduce food quality (Kaspari and Welti 2024; Sardans et al. 2012;

Stiling and Cornelissen 2007; Welti et al. 2020). For example, several studies have documented increases in phenolic compounds under elevated CO_2 (Johnson et al. 2022; Robinson et al. 2012). In other words, plants grow faster and larger, but they can also become tougher, less palatable, and less nutritious. However, the extent to which this happens depends further on plant functional strategy. C4 plants remain relatively unresponsive to elevated CO_2, while C3 plants often show marked growth increases. Legumes may benefit even more strongly, channelling excess carbon to their nitrogen-fixing symbionts, which not only boosts legume growth but also enriches ecosystem nitrogen availability (Rogers et al. 2009). Over longer timescales, this can lead to changes in plant composition, favouring responsive C3 species and legumes while reducing the relative abundance of less responsive groups.

Taken together, the implications of elevated CO_2 for herbivores are complex: More biomass may mean more resources, but plant tissues are often of poorer quality and might not meet the stoichiometric demands of herbivores (Kaspari and Welti 2024), ultimately dampening any changes in herbivore abundance.

5.3.3 Land-Use Intensification Through Nitrogen Fertilisation, Mowing, Grazing, and Pesticide Use

Land-use intensification through grazing, mowing, pesticide use, and nitrogen fertilisation directly affects insect herbivores in several ways. Large vertebrate grazers, such as livestock and wild ungulates, may trample or incidentally ingest invertebrates when feeding, causing direct mortality and reducing local insect abundance (Gómez and González-Megías 2002; van Klink et al. 2014). However, it is mowing that has the most detrimental impact on invertebrate herbivore abundance and diversity, compared to other management practices such as grazing or fertilisation (Chisté et al. 2016; Humbert et al. 2010; Proske et al. 2022). Although mowing techniques differ mechanically, all represent severe disturbances for ground-dwelling and low-flying arthropods (Berger et al. 2024; Humbert et al. 2010). Rotary mowers and mulchers usually cause higher mortality than bar mowers (Humbert et al. 2010; Steidle et al. 2022), but such differences are often outweighed by the disturbance itself and by subsequent operations. The removal and processing of hay (e.g. tedding, raking, baling) can eliminate insects and their overwintering stages remaining in the cut biomass (Hecker et al. 2022). Even technically "insect-friendly" mowing practices therefore result in substantial arthropod losses. In addition, heavy machinery can compact soil and reduce shelter, forcing emigration from mown areas (Berger et al. 2024). Hence, mowing, regardless of technique, acts as a major physical disturbance that directly removes both organisms and microhabitats from grassland systems.

The use of pesticides also directly harms insect herbivores by suppressing their growth, survival, and reproductive success, with insecticides disrupting neural and metabolic pathways essential for feeding and development. Although most research on non-target pesticide effects has focused on pollinators (Woodcock et al. 2016), recent syntheses show that impacts are widespread across taxa and likely extend to herbivorous groups as well (Wan et al. 2025). These large-scale analyses suggest that pesticide exposure is an important contributor to global insect declines, with the potential to impair herbivore performance and population persistence in both managed and natural ecosystems (Geiger et al. 2010; Quandahor et al. 2024). By contrast, nitrogen enrichment has no direct physiological effects on herbivores; its impacts are entirely indirect, mediated through changes in plant growth, chemistry, and community composition.

Land-use practices such as nitrogen fertilisation, grazing, mowing, and pesticide application also have strong indirect effects on insect herbivores by reshaping the composition, biomass, and diversity of plant communities. Nitrogen enrichment is particularly influential. In the short term, it increases foliar nitrogen concentrations, thereby enhancing the nutritional quality of plant tissue (Ebeling et al. 2022) while simultaneously increasing plant biomass production. Over longer time scales, nitrogen enrichment intensifies light competition and shifts plant communities toward dominance by fast-growing, palatable species, at the expense of slower-growing taxa, thereby reducing plant diversity and evenness (Coley et al. 1985; Hautier et al. 2009; Lind et al. 2013; Stevens et al. 2004). Heavy grazing, by contrast, tends to favour slow-growing, well-defended plant species that are less palatable, again resulting in reduced diversity but very different community composition (Milchunas et al. 1988; Olff and Ritchie 1998). Grazing and trampling also alter the physical habitat that small herbivores experience, for example, by compressing litter layers, removing vegetation cover, or modifying soil microclimate conditions such as temperature and moisture (Huntzinger et al. 2008; Suominen et al. 1999). Such changes expose insects to desiccation, thermal stress, and predation and may also disturb oviposition sites or remove critical microhabitats for ground- and foliage-dwelling insects (van Klink et al. 2014). Although these indirect effects via trampling are often negative, their magnitude depends on grazer density, timing, and mobility and can be locally offset when grazing creates spatial heterogeneity or disturbance refugia that some insect taxa can exploit. High rates of grazing and mowing also clearly reduce the amount of plant biomass available for other consumers, such as insects. By comparison, moderate levels of grazing and mowing can sustain higher plant diversity, preventing competitive exclusion and generating fine-scale heterogeneity in vegetation structure (Jerrentrup et al. 2014; Panassiti et al. 2025). At the other end of the spectrum, the complete absence of mowing or grazing also undermines plant diversity, as tall, competitive species monopolise resources and homogenise the vegetation (Hempel et al. 2025).

Although nitrogen inputs almost invariably reduce plant diversity by promoting fast-growing species, they simultaneously increase bottom-up resources for herbivores through higher foliar nitrogen and greater biomass production. The consequences for insect herbivores are therefore complex: Abundance may rise due to improved food quality (Chase et al. 2000; Ebeling et al. 2022; Lind et al. 2017), but diversity is likely to decline, and community composition often shifts toward dominance by species adapted to exploit fast, palatable plants (Bobbink et al. 2010; De Schrijver et al. 2011; Isbell et al. 2013). In contrast, the effects of grazing and mowing are often non-linear: Very high intensities reduce plant diversity and remove the biomass that would otherwise support herbivores, leading to declines in insect herbivore abundance and diversity. However, moderate grazing and mowing regimes maintain diverse plant assemblages and generate structural heterogeneity, supporting a richer and more balanced herbivore community. No grazing or mowing is also detrimental for many herbivores as it leads to a loss of plant diversity and therefore a loss of the herbivores that depend on varied plant resources (Hempel et al. 2025).

5.3.4 Invasive Plant Species

Invasive plant species can have direct negative effects on herbivorous insects, because they replace the native plants that the herbivores are adapted to (Skinner et al. 2026). Many invasive plants are poorly utilised by specialist herbivores due to mismatches in plant chemistry, morphology, or phenology, and invasion therefore often reduces the availability of suitable food resources even when overall plant biomass increases (Burghardt et al. 2010; Grandez-Rios et al. 2015; Harvey et al. 2010a; Tallamy and Shropshire 2009). Invasive plants may also dilute host plant signals or physically occupy space that would otherwise be used for feeding, oviposition, or shelter (Harvey et al. 2010b). Such direct effects tend to be most pronounced for specialist herbivores with narrow dietary niches, whereas some generalist species may be able to exploit invasive plants to a limited degree.

Beyond these direct host-related effects, invasive plants also influence herbivores indirectly by altering plant community composition, diversity, and structure (Powell et al. 2011). By competitively displacing native species, invasive plants often homogenise vegetation and reduce the diversity of niches and microhabitats available to insects (Daru et al. 2021; Thakur et al. 2025). These community shifts frequently cascade into changes in food-web structure, nutrient cycling, and soil properties, for instance, through modified litter chemistry or altered microbial communities (Liao et al. 2008), which can further constrain herbivore performance. The magnitude and direction of indirect effects depend on the identity of the invasive species and the vulnerability of the invaded community (Skinner et al. 2026; Thakur et al. 2025). However, in most cases, the net outcome for native insect herbivores is negative (Skinner et al. 2026), with declines in diversity and shifts toward highly generalist or disturbance-tolerant taxa.

5.4 Variation in Direct and Indirect Effects Across Managed and Unmanaged Ecosystems

The importance of direct versus indirect global change effects depends strongly on ecosystem context. In natural and semi-natural ecosystems, global change can affect herbivores both directly and indirectly through changes in plant communities. Because plant species composition, diversity, and biomass are free to respond to environmental drivers such as warming, altered precipitation, or nutrient inputs, indirect pathways mediated by vegetation change may be equally or more important than direct physiological effects on herbivores. For example, warming may increase herbivore abundance by enhancing plant biomass or nutritional quality, but these effects could also be dampened if warming and increased herbivory favours slow-growing, well-defended plant species, leading to reduced forage quality. Because such community shifts occur slowly, indirect effects may emerge only after substantial temporal lags (see Chap. 4).

By contrast, in highly managed systems such as intensive agriculture, forestry, or seeded grasslands, plant communities are actively controlled by humans, and species composition cannot respond freely to climate or nutrient changes or increases in herbivory. The conversion of natural or semi-natural habitats to such simplified systems therefore imposes immediate direct pressures on herbivores, including mowing, pesticide exposure, and habitat disturbance while simultaneously constraining indirect pathways mediated by vegetation change. Under these conditions, global change drivers are expected to act primarily through direct physiological or demographic effects on herbivores, for example, by increasing metabolic demands in herbivores, potentially leading to yield losses under warming (Deutsch et al. 2018). Overall, the balance between direct and indirect global change effects varies strongly across ecosystems, depending on management intensity and the degree to which plant communities are able to reorganise in response to environmental change and altered herbivore pressure.

5.5 Predicting the Net Outcome of Direct and Indirect Effects on Herbivores

Herbivore responses to different global change drivers emerge from the interplay between direct physiological or demographic effects and indirect effects mediated through plant communities and associated food webs. Warming typically enhances invertebrate metabolic rate, growth, and reproduction, whereas land-use practices such as mowing, trampling, or pesticide exposure tend to reduce survival. Invasive plants directly diminish resource quality when they replace native hosts with species that local herbivores cannot efficiently exploit. However, these direct effects can be modified, i.e. strengthened or counteracted, by indirect plant-mediated pathways. For instance, light grazing may support higher plant diversity, which could increase insect herbivore diversity and abundance.

Each global change driver is also likely to cause a number of different indirect effects that may enhance or dampen each other's effects. For example, nitrogen enrichment can increase plant biomass and foliar nitrogen concentrations, which should benefit insect herbivore; however, it also reduces plant diversity, which might negate some of those positive effects. In addition, elevated CO_2, drought, and some invasive plant species reduce food quality by increasing carbon-based defences, lowering leaf nitrogen, or altering secondary metabolites, thereby suppressing herbivore performance despite potentially leading to higher plant biomass (de Sassi and Tylianakis 2012; Hagvar and Klanderud 2009). In contrast, invasive species can indirectly reduce herbivore diversity by homogenising plant communities and altering soil processes, which may enhance their direct effects caused by replacement of native hosts.

Consequently, global change drivers rarely produce uniformly positive or negative outcomes. Instead, the net response depends on the relative strength and direction of different pathways acting simultaneously on plants and herbivores. Trait differences among herbivores—such as feeding specialisation, mobility, thermal sensitivity, and vulnerability to predators—further modulate these responses and help explain why generalists sometimes increase while specialist taxa decline. Because direct and indirect pathways often act in opposing directions, accurate forecasts of herbivore responses to global change will require explicitly jointly assessing and teasing apart *direct* effects and *indirect* plant-mediated effects and understanding their interactions across trophic levels.

5.6 Disentangling Direct and Indirect Effects of Multiple Global Change Drivers

Disentangling direct from indirect effects of global change on herbivores and consequently their interactions with plants is essential for predicting ecosystem responses. Yet, most large-scale studies quantify only the overall relationship between environmental gradients or global change factors and herbivore communities (Liu et al. 2024), overlooking the relative importance of direct versus plant-mediated effects. This limitation hampers our ability to accurately forecast how herbivory, plant diversity, or ecosystem functioning will respond to future environmental change. One way of addressing these complexities and disentangling direct effects and indirect effects is to combine coordinated measurements of plant and herbivore data across broad climatic gradients with structural equation models (SEMs, Box 5.1). SEMs allow simultaneous testing of multiple causal pathways, making it possible to quantify how climate and land-use drivers influence herbivores both directly and indirectly via changes in plant diversity, functional traits, plant quality, and productivity (e.g. Cappelli et al. 2020). By comparing the strength and direction of these pathways, SEMs help reveal the relative importance of direct versus plant-mediated effects, providing a mechanistic understanding of how global change shapes herbivore dynamics across scales (Box 5.1).

However, SEMs applied to observational data can never fully reveal causality; for this, we need experiments. Disentangling the relative importance of direct versus indirect effects of global change on plant–herbivore interactions can be done with factorial experiments that simultaneously manipulate plant community characteristics and global change drivers (Fig. 5.2). Such experiments can cross manipulations of climate or nutrient enrichment with manipulations of plant diversity or functional trait composition. Several large-scale biodiversity and global change experiments have adopted this approach (Craven et al. 2016, Fig. 5.2). For example, the PaNDiv experiment manipulates plant diversity and functional composition together with nitrogen addition (Pichon et al. 2020); the Jena Experiment manipulates plant diversity together with drought and nitrogen addition (Vogel et al. 2012;

Weigelt et al. 2009); the BioCON experiment combines plant diversity and nitrogen and CO_2 enrichment treatments (Isbell et al. 2013); the BioClive experiment manipulates plant diversity and future precipitation scenarios (Hautier et al. 2024); the ORPHEE experiment manipulates tree diversity and functional composition and water availability through irrigation (Castagneyrol et al. 2013); and, finally, the BAC experiment investigates the joint effects of plant diversity and experimental warming (Cowles et al. 2016). By quantifying invertebrate abundance or herbivory across these treatment combinations, such experiments can estimate the relative contributions of various direct and plant-mediated pathways, offering causal insights into how global change affects herbivore communities across biodiversity gradients. Specifically, direct effects can be inferred when herbivore responses occur independently of plant community manipulations, e.g. when drought affects herbivores independently of plant species richness. Indirect effects are likely to be more important when direct effects of drought are absent, and responses depend on changes in plant diversity or composition. Analytical approaches such as variance partitioning or structural equation modelling can then be applied to quantify how much of the variation in herbivore responses is explained by direct versus plant-mediated pathways, providing a mechanistic understanding of global change impacts. Moving forward, integrating such comparative and experimental approaches will be key to developing a mechanistic and predictive understanding of how global change will affect herbivores and ecosystem function in the future.

Box 5.1: The Bug-Network, a Comparative Approach to Disentangle Direct and Indirect Effects of Global Change on Herbivory and Pathogen Damage

The comparative component of a novel global network aiming to better understand the impact of invertebrate herbivores and pathogens on plant communities and ecosystem functioning, the Bug Network (BugNet; https://bug-net.org), provides a unique global framework to disentangle direct and indirect effects of global change on plant–consumer interactions. Researchers collected detailed information on plant community characteristics (i.e. diversity, composition, biomass, functional traits, nutritive value, chemical defences) and herbivore and pathogen damage across 114 sites worldwide, spanning a wide range of climatic and environmental contexts. This dataset can then be analysed using a multi-level structural equation modelling framework, which enables the quantification of both direct climatic influences and indirect pathways mediated by plant community responses, from local to global scales. This combination of globally coordinated field sampling and advanced causal modelling represents a powerful approach to identify the mechanisms linking climate, plants, and their herbivores. Such comparative, data-intensive frameworks are one example for moving beyond correlative patterns and toward a mechanistic understanding of how global change reshapes trophic interactions worldwide. Image designed by Julien L. Bota.

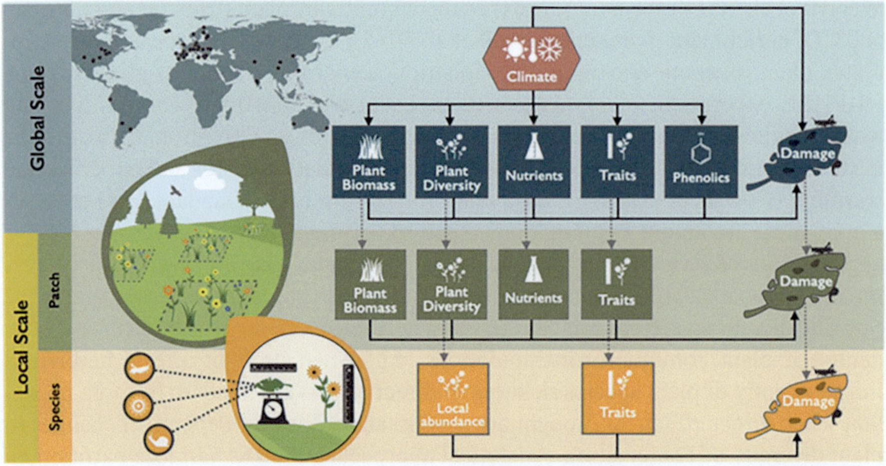

Fig. 5.2 Examples of experiments that cross manipulations of climate or nutrient enrichment with manipulations of plant diversity or functional composition. (**a**) The Jena Experiment manipulates plant diversity together with drought and nitrogen addition (Foto: The Jena Experiment); (**b**) the PaNDiv experiment manipulates plant diversity and functional composition together with nitrogen addition and pathogen exclusion (Foto: H. Vincent); (**c**) the BioCON experiment combines plant diversity and nitrogen and CO_2 enrichment treatments (Foto: M. Liang); (**d**) the ORPHEE experiment manipulates tree diversity and functional composition and water availability through irrigation (Foto: B. Issenhuth)

5.7 Herbivore Effects on Plant Communities and Implications Under Global Change

While the preceding sections highlighted how global change effects on plants can cascade to influence herbivore communities, it is important to remember that herbivores themselves also exert strong top-down control on plants. In many cases, they can modify or even reverse plant responses to environmental change (Borer et al. 2014; Lemoine et al. 2017; Seabloom et al. 2017), sometimes involving feedbacks. By consuming plant biomass, herbivores directly reduce primary productivity (Crawley 1997). Their selective feeding also influences which plant species thrive in a community. When herbivores mostly attack dominant plant species, they can open space for others and help maintain plant diversity by holding the dominant species in check (Pacala and Crawley 1992). In contrast, when herbivores prefer rare species, they can reduce plant diversity (Allan and Crawley 2011). Herbivores frequently drive shifts in plant functional composition too. Their feeding tends to favour slow-growing, well-defended plant species and disadvantage fast-growing, palatable ones (Cappelli et al. 2020; Kempel et al. 2015; Lind et al. 2013). Although these patterns are well documented, we still lack a clear understanding of the overall strength, context dependency, and generality of herbivore impacts, especially under global change scenarios (Jia et al. 2018; Jiang et al. 2025).

Even so, a substantial foundation of research already exists. Many studies have explored how herbivores respond to environmental gradients and to individual global change drivers. Recent work—through targeted experiments such as herbivore exclusions across abiotic gradients (such as within the Bug-Network, Kempel et al. 2025) and multifactorial global change manipulations (Smith et al. 2024; Stevens et al. 2004)—offers a promising path toward more predictive insight into herbivore dynamics under changing conditions. This is crucial, because shifts in herbivore communities can feedback to plant communities and either amplify or dampen global change effects on ecosystems, similar to what we observe in climate change research, where warming creates change (i.e. thawing permafrost) which then either slows down or speeds up warming (IPCC 2023). Such plant-herbivore feedbacks might be important in accurately predicting global change effects on ecosystems and overlooking these feedbacks risks misleading predictions or interpretations. For instance, incorporating feedbacks might lead to altered predictions for global carbon models: While warming might be expected to enhance crop yields or carbon sequestration through increases in plant productivity, these gains can also fuel higher herbivore populations, reducing or even reversing the expected benefits (Bebber et al. 2013; Deutsch et al. 2018; Pringle et al. 2023). Accordingly, several studies already show that climate-driven boosts in plant growth can be offset by intensified pathogen infection (Chaloner et al. 2021) and insect herbivory (Couture et al. 2015). Despite growing evidence, we still lack a comprehensive understanding of how multiple global change drivers jointly shape plant–herbivore interactions across scales and ecosystems. Strengthening our understanding of plant–herbivore interactions will therefore be key for forecasting plant community and ecosystem responses in the future.

5.8 Implications for Ecosystem Management

A better understanding of the direct and indirect pathways through which global change drivers affect herbivores may also improve ecosystem management. Many common management practices, such as mowing and grazing, impose strong direct pressures on invertebrate herbivores by directly killing herbivores. At the same time, these practices can indirectly benefit herbivores by promoting more diverse and functionally heterogeneous plant communities or by increasing the proportion of young, nutritious plant tissue. Designing management strategies that minimse negative direct impacts while retaining or enhancing positive plant-mediated effects is therefore essential. For instance, incorporating insect refuges—such as uncut field margins, flower strips, or mosaic mowing regimes—can buffer herbivore populations from the immediate disturbance of mowing or grazing while still allowing managers to maintain open, productive systems (Herzon et al. 2026; Révész et al. 2025).

Management approaches that modify plant traits can also help buffer herbivore responses to global change. For example, in agricultural systems, the use of pest-resistant or stress-tolerant crop varieties may counterbalance the expected increase in herbivore pressure driven by direct warming effects (Mitchell et al. 2016; Sun et al. 2025). Heat-enhanced herbivore metabolism and feeding rates can substantially boost crop damage, but resistant cultivars—with traits such as tougher tissues, secondary metabolites, or inducible defences—can reduce vulnerability and stabilise yields under warming (Sun et al. 2025).

Similarly, management actions that enhance or restore plant diversity may help counteract some of the negative indirect effects of nitrogen enrichment, pesticides, or land-use intensification. Diverse plant communities tend to support more stable and resilient herbivore assemblages, partly because they provide a broader range of nutritional resources and microhabitats. Practices such as reducing fertilisation intensity, diversifying crop rotations, adopting low-input agriculture, or restoring heterogeneous grassland structures can therefore strengthen stabilising plant–herbivore linkages and improve ecosystem functioning (Herzon et al. 2026). Integrating knowledge of both direct and indirect pathways into management planning will be crucial for maintaining biodiversity, promoting ecosystem resilience, and securing ecosystem services under ongoing global change.

References

Ainsworth EA, Long SP (2004) What have we learned from 15 years of free-air CO2 enrichment (FACE)? A meta-analytic review of the responses of photosynthesis, canopy properties and plant production to rising CO2. New Phytol 165(2):351–372. https://doi.org/10.1111/j.1469-8137.2004.01224.x

Ali JG, Agrawal AA (2012) Specialist versus generalist insect herbivores and plant defense. Trends Plant Sci 17(5):293–302. https://doi.org/10.1016/j.tplants.2012.02.006

Allan E, Crawley MJ (2011) Contrasting effects of insect and molluscan herbivores on plant diversity in a long-term field experiment: invertebrate herbivores and biodiversity. Ecol Lett 14(12):1246–1253. https://doi.org/10.1111/j.1461-0248.2011.01694.x

Allan E, Manning P, Alt F, Binkenstein J, Blaser S, Blüthgen N, Böhm S, Grassein F, Hölzel N, Klaus VH, Kleinebecker T, Morris EK, Oelmann Y, Prati D, Renner SC, Rillig MC, Schaefer M, Schloter M, Schmitt B et al (2015) Land use intensification alters ecosystem multifunctionality via loss of biodiversity and changes to functional composition. Ecol Lett 18(8):834–843. https://doi.org/10.1111/ele.12469

Ampoorter E, Barbaro L, Jactel H, Baeten L, Boberg J, Carnol M, Castagneyrol B, Charbonnier Y, Dawud SM, Deconchat M, Smedt PD, Wandeler HD, Guyot V, Hättenschwiler S, Joly F-X, Koricheva J, Milligan H, Muys B, Nguyen D et al (2020) Tree diversity is key for promoting the diversity and abundance of forest-associated taxa in Europe. Oikos 129(2):133–146. https://doi.org/10.1111/oik.06290

Aunapuu M, Dahlgren J, Oksanen T, Grellmann D, Oksanen L, Olofsson J, Rammul Ü, Schneider M, Johansen B, Hygen HO (2008) Spatial patterns and dynamic responses of Arctic food webs corroborate the exploitation ecosystems hypothesis (EEH). Am Nat 171(2):249–262. https://doi.org/10.1086/524951

Bagchi R, Gallery RE, Gripenberg S, Gurr SJ, Narayan L, Addis CE, Freckleton RP, Lewis OT (2014) Pathogens and insect herbivores drive rainforest plant diversity and composition. Nature 506(7486):85–88. https://doi.org/10.1038/nature12911

Bale JS, Masters GJ, Hodkinson ID, Awmack C, Bezemer TM, Brown VK, Butterfield J, Buse A, Coulson JC, Farrar J, Good JEG, Harrington R, Hartley S, Jones TH, Lindroth RL, Press MC, Symrnioudis I, Watt AD, Whittaker JB (2002) Herbivory in global climate change research: direct effects of rising temperature on insect herbivores. Glob Chang Biol 8(1):1–16. https://doi.org/10.1046/j.1365-2486.2002.00451.x

Barbosa P, Hines J, Kaplan I, Martinson H, Szczepaniec A, Szendrei Z (2009) Associational resistance and associational susceptibility: having right or wrong neighbors. Annu Rev Ecol Evol Syst 40(1):1–20. https://doi.org/10.1146/annurev.ecolsys.110308.120242

Barnes AD, Scherber C, Brose U, Borer ET, Ebeling A, Gauzens B, Giling DP, Hines J, Isbell F, Ristok C, Tilman D, Weisser WW, Eisenhauer N (2020) Biodiversity enhances the multitrophic control of arthropod herbivory. Sci Adv 6(45):eabb6603. https://doi.org/10.1126/sciadv.abb6603

Bebber DP, Ramotowski MAT, Gurr SJ (2013) Crop pests and pathogens move polewards in a warming world. Nat Clim Chang 3(11):985–988. https://doi.org/10.1038/nclimate1990

Berger JL, Staab M, Hartlieb M, Simons NK, Wells K, Gossner MM, Vogt J, Achury R, Seibold S, Hemp A, Weisser WW, Blüthgen N (2024) The day after mowing: time and type of mowing influence grassland arthropods. Ecol Appl 34(6):e3022. https://doi.org/10.1002/eap.3022

Bernays EA, Bright KL, Gonzalez N, Angel J (1994) Dietary mixing in a generalist herbivore: tests of two hypotheses. Ecology 75(7):1997–2006. https://doi.org/10.2307/1941604

Bobbink R, Hicks K, Galloway J, Spranger T, Alkemade R, Ashmore M, Bustamante M, Cinderby S, Davidson E, Dentener F, Emmett B, Erisman J-W, Fenn M, Gilliam F, Nordin A, Pardo L, De Vries W (2010) Global assessment of nitrogen deposition effects on terrestrial plant diversity: a synthesis. Ecol Appl 20(1):30–59. https://doi.org/10.1890/08-1140.1

Borer ET, Seabloom EW, Tilman D (2012) Plant diversity controls arthropod biomass and temporal stability. Ecol Lett 15(12):1457–1464. https://doi.org/10.1111/ele.12006

Borer ET, Seabloom EW, Gruner DS, Harpole WS, Hillebrand H, Lind EM, Adler PB, Alberti J, Anderson TM, Bakker JD, Biederman L, Blumenthal D, Brown CS, Brudvig LA, Buckley YM, Cadotte M, Chu C, Cleland EE, Crawley MJ et al (2014) Herbivores and nutrients control grassland plant diversity via light limitation. Nature 508(7497):517–520. https://doi.org/10.1038/nature13144

Brändle M, Amarell U, Auge H, Klotz S, Brandl R (2001) Plant and insect diversity along a pollution gradient: understanding species richness across trophic levels. Biodivers Conserv 10(9):1497–1511. https://doi.org/10.1023/A:1011815325503

Brown JH, Gillooly JF, Allen AP, Savage VM, West GB (2004) Toward a metabolic theory of ecology. Ecology 85(7):1771–1789. https://doi.org/10.1890/03-9000

Burghardt KT, Tallamy DW, Philips C, Shropshire KJ (2010) Non-native plants reduce abundance, richness, and host specialization in lepidopteran communities. Ecosphere 1(5):1–22. https://doi.org/10.1890/es10-00032.1

Cappelli SL, Pichon NA, Kempel A, Allan E (2020) Sick plants in grassland communities: a growth-defense trade-off is the main driver of fungal pathogen abundance. Ecol Lett 23(9):1349–1359. https://doi.org/10.1111/ele.13537

Carson WP, Root RB (2000) Herbivory and plant species coexistence: community regulation by an outbreaking phytophagous insect. Ecol Monogr 70(1):73–99

Castagneyrol B, Giffard B, Péré C, Jactel H (2013) Plant apparency, an overlooked driver of associational resistance to insect herbivory. J Ecol 101(2):418–429. https://doi.org/10.1111/1365-2745.12055

Chaloner TM, Gurr SJ, Bebber DP (2021) Plant pathogen infection risk tracks global crop yields under climate change. Nat Clim Chang 11(8):710–715. https://doi.org/10.1038/s41558-021-01104-8

Chase JM, Leibold MA, Downing AL, Shurin JB (2000) The effects of productivity, herbivory, and plant species turnover in grassland food webs. Ecology 81(9):2485–2497

Chen C, Harvey JA, Biere A, Gols R (2019) Rain downpours affect survival and development of insect herbivores: the specter of climate change? Ecology 100(11):e02819. https://doi.org/10.1002/ecy.2819

Chesson P (2000) Mechanisms of maintenance of species diversity. Annu Rev Ecol Syst 31(1):343–366. https://doi.org/10.1146/annurev.ecolsys.31.1.343

Chisté MN, Mody K, Gossner MM, Simons NK, Köhler G, Weisser WW, Blüthgen N (2016) Losers, winners, and opportunists: how grassland land-use intensity affects orthopteran communities. Ecosphere 7(11):e01545. https://doi.org/10.1002/ecs2.1545

Coakley SM, Scherm H, Chakraborty S (1999) Climate change and plant disease management. Annu Rev Phytopathol 37(1):399–426. https://doi.org/10.1146/annurev.phyto.37.1.399

Coley PD, Kursar TA (2014) On tropical forests and their pests. Science 343(6166):35–36. https://doi.org/10.1126/science.1248110

Coley PD, Bryant JP, Chapin FS (1985) Resource availability and plant Antiherbivore defense. Science 230(4728):895–899. https://doi.org/10.1126/science.230.4728.895

Cornelissen T (2011) Climate change and its effects on terrestrial insects and herbivory patterns. Neotrop Entomol 40(2):155–163. https://doi.org/10.1590/s1519-566x2011000200001

Couture JJ, Meehan TD, Kruger EL, Lindroth RL (2015) Insect herbivory alters impact of atmospheric change on northern temperate forests. Nat Plants 1(3):15016. https://doi.org/10.1038/nplants.2015.16

Cowles JM, Wragg PD, Wright AJ, Powers JS, Tilman D (2016) Shifting grassland plant community structure drives positive interactive effects of warming and diversity on aboveground net primary productivity. Glob Chang Biol 22(2):741–749. https://doi.org/10.1111/gcb.13111

Craven D, Isbell F, Manning P, Connolly J, Bruelheide H, Ebeling A, Roscher C, van Ruijven J, Weigelt A, Wilsey B, Beierkuhnlein C, de Luca E, Griffin JN, Hautier Y, Hector A, Jentsch A, Kreyling J, Lanta V, Loreau M et al (2016) Plant diversity effects on grassland productivity are robust to both nutrient enrichment and drought. Philos Trans R Soc B Biol Sci 371(1694):20150277. https://doi.org/10.1098/rstb.2015.0277

Crawley MJ (1983) Herbivory: the dynamics of plant animal interactions. Blackwell Scientific

Crawley M (1997) Plant ecology. Wiley. https://books.google.ch/books?id=TnI9FBYNR3QC

Daru BH, Davies TJ, Willis CG, Meineke EK, Ronk A, Zobel M, Pärtel M, Antonelli A, Davis CC (2021) Widespread homogenization of plant communities in the Anthropocene. Nat Commun 12(1):6983. https://doi.org/10.1038/s41467-021-27186-8

de Sassi C, Tylianakis JM (2012) Climate change disproportionately increases herbivore over plant or parasitoid biomass. PLoS One 7(7):e40557. https://doi.org/10.1371/journal.pone.0040557

De Schrijver A, De Frenne P, Ampoorter E, Van Nevel L, Demey A, Wuyts K, Verheyen K (2011) Cumulative nitrogen input drives species loss in terrestrial ecosystems. Glob Ecol Biogeogr 20(6):803–816. https://doi.org/10.1111/j.1466-8238.2011.00652.x

Deraison H, Badenhausser I, Loeuille N, Scherber C, Gross N (2015) Functional trait diversity across trophic levels determines herbivore impact on plant community biomass. Ecol Lett 18(12):1346–1355. https://doi.org/10.1111/ele.12529

Deutsch CA, Tewksbury JJ, Tigchelaar M, Battisti DS, Merrill SC, Huey RB, Naylor RL (2018) Increase in crop losses to insect pests in a warming climate. Science 361(6405):916–919. https://doi.org/10.1126/science.aat3466

Díaz S, Kattge J, Cornelissen JHC, Wright IJ, Lavorel S, Dray S, Reu B, Kleyer M, Wirth C, Colin Prentice I, Garnier E, Bönisch G, Westoby M, Poorter H, Reich PB, Moles AT, Dickie J, Gillison AN, Zanne AE et al (2016) The global spectrum of plant form and function. Nature 529(7585):167–171. https://doi.org/10.1038/nature16489

Dobson KC, Zarnetske PL (2025) A global meta-analysis of passive experimental warming effects on plant traits and community properties. Glob Chang Biol 31(6):e70306. https://doi.org/10.1111/gcb.70306

Ebeling A, Meyer ST, Abbas M, Eisenhauer N, Hillebrand H, Lange M, Scherber C, Vogel A, Weigelt A, Weisser WW (2014) Plant diversity impacts decomposition and herbivory via changes in aboveground arthropods. PLoS One 9(9):e106529. https://doi.org/10.1371/journal.pone.0106529

Ebeling A, Strauss AT, Adler PB, Arnillas CA, Barrio IC, Biederman LA, Borer ET, Bugalho MN, Caldeira MC, Cadotte MW, Daleo P, Eisenhauer N, Eskelinen A, Fay PA, Firn J, Graff P, Hagenah N, Haider S, Komatsu KJ et al (2022) Nutrient enrichment increases invertebrate herbivory and pathogen damage in grasslands. J Ecol 110(2):327–339. https://doi.org/10.1111/1365-2745.13801

Feeny P (1976) Plant apparency and chemical defense. In: Wallace JW, Mansell RL (eds) Biochemical interaction between plants and insects. Springer US, pp 1–40. https://doi.org/10.1007/978-1-4684-2646-5_1

Forister ML, Novotny V, Panorska AK, Baje L, Basset Y, Butterill PT, Cizek L, Coley PD, Dem F, Diniz IR, Drozd P, Fox M, Glassmire AE, Hazen R, Hrcek J, Jahner JP, Kaman O, Kozubowski TJ, Kursar TA et al (2015) The global distribution of diet breadth in insect herbivores. Proc Natl Acad Sci 112(2):442–447. https://doi.org/10.1073/pnas.1423042112

Franco ALC, Gherardi LA, de Tomasel CM, Andriuzzi WS, Ankrom KE, Shaw EA, Bach EM, Sala OE, Wall DH (2019) Drought suppresses soil predators and promotes root herbivores in Mesic, but not in xeric grasslands. Proc Natl Acad Sci 116(26):12883–12888. https://doi.org/10.1073/pnas.1900572116

Franklin J, Serra-Diaz JM, Syphard AD, Regan HM (2016) Global change and terrestrial plant community dynamics. Proc Natl Acad Sci 113(14):3725–3734. https://doi.org/10.1073/pnas.1519911113

Fretwell SD (1987) Food chain dynamics: the central theory of ecology? Oikos 50(3):291. https://doi.org/10.2307/3565489

Geiger F, Bengtsson J, Berendse F, Weisser WW, Emmerson M, Morales MB, Ceryngier P, Liira J, Tscharntke T, Winqvist C, Eggers S, Bommarco R, Pärt T, Bretagnolle V, Plantegenest M, Clement LW, Dennis C, Palmer C, Oñate JJ et al (2010) Persistent negative effects of pesticides on biodiversity and biological control potential on European farmland. Basic Appl Ecol 11(2):97–105. https://doi.org/10.1016/j.baae.2009.12.001

Gilman SE, Urban MC, Tewksbury J, Gilchrist GW, Holt RD (2010) A framework for community interactions under climate change. Trends Ecol Evol 25(6):325–331. https://doi.org/10.1016/j.tree.2010.03.002

Giolai M, Laine A-L (2024) A trade-off between investment in molecular defense repertoires and growth in plants. Science 386(6722):677–680. https://doi.org/10.1126/science.adn2779

Gómez JM, González-Megías A (2002) Asymmetrical interactions between ungulates and phytophagous insects: being different matters. Ecology 83(1):203–211

Grace JB, Anderson TM, Seabloom EW, Borer ET, Adler PB, Harpole WS, Hautier Y, Hillebrand H, Lind EM, Pärtel M, Bakker JD, Buckley YM, Crawley MJ, Damschen EI, Davies KF, Fay PA, Firn J, Gruner DS, Hector A et al (2016) Integrative modelling reveals mechanisms linking productivity and plant species richness. Nature 529(7586):390–393. https://doi.org/10.1038/nature16524

Grandez-Rios JM, Lima Bergamini L, Santos de Araújo W, Villalobos F, Almeida-Neto M (2015) The effect of host-plant phylogenetic isolation on species richness, composition and specialization of insect herbivores: a comparison between native and exotic hosts. PLoS One 10(9):e0138031. https://doi.org/10.1371/journal.pone.0138031

Haddad NM, Haarstad J, Tilman D (2000) The effects of long-term nitrogen loading on grassland insect communities. Oecologia 124(1):73–84. https://doi.org/10.1007/s004420050026

Haddad NM, Tilman D, Haarstad J, Ritchie M, Knops JMH (2001) Contrasting effects of plant richness and composition on insect communities: a field experiment. Am Nat 158(1):17. https://doi.org/10.2307/3078895

Hagvar S, Klanderud K (2009) Effect of simulated environmental change on alpine soil arthropods. Glob Chang Biol 15(12):2972–2980. https://doi.org/10.1111/j.1365-2486.2009.01926.x

Hairston NG, Smith FE, Slobodkin LB (1960) Community structure, population control, and competition. Am Nat 94(879):421–425. https://doi.org/10.1086/282146

Halliday FW, Heckman RW, Wilfahrt PA, Mitchell CE (2017) A multivariate test of disease risk reveals conditions leading to disease amplification. Proc R Soc B Biol Sci 284(1865):20171340. https://doi.org/10.1098/rspb.2017.1340

Hamann E, Blevins C, Franks SJ, Jameel MI, Anderson JT (2021) Climate change alters plant–herbivore interactions. New Phytol 229(4):1894–1910. https://doi.org/10.1111/nph.17036

Harvey JA, Biere A, Fortuna T, Vet LEM, Engelkes T, Morriën E, Gols R, Verhoeven K, Vogel H, Macel M, Heidel-Fischer HM, Schramm K, van der Putten WH (2010a) Ecological fits, mis-fits and lotteries involving insect herbivores on the invasive plant, Bunias orientalis. Biol Invasions 12(9):3045–3059. https://doi.org/10.1007/s10530-010-9696-9

Harvey JA, Bukovinszky T, van der Putten WH (2010b) Interactions between invasive plants and insect herbivores: a plea for a multitrophic perspective. Biol Conserv 143(10):2251–2259. https://doi.org/10.1016/j.biocon.2010.03.004

Hautier Y, Niklaus PA, Hector A (2009) Competition for light causes plant biodiversity loss after eutrophication. Science 324(5927):636–638. https://doi.org/10.1126/science.1169640

Hautier Y, Barry KE, Hefting MM, van Kuijk M, Pos ET, Verduyn BP, Johannes R, Kowalchuk GA, Soons MB (2024) The biodiversity and climate variability experiment (BioCliVE): quantifying the role of biodiversity in buffering ecosystems against climatic variability. Res Ideas Outcomes 10:e133454. https://doi.org/10.3897/rio.10.e133454

Hecker LP, Wätzold F, Yang X, Birkhofer K (2022) Squeeze it or leave it? An ecological-economic assessment of the impact of mower conditioners on arthropod populations in grassland. J Insect Conserv 26(3):463–475. https://doi.org/10.1007/s10841-022-00392-5

Hempel S, Herzog F, Batáry P, Öckinger E, Knop E (2025) The impact of abandonment and intensification on the biodiversity of agriculturally marginal grasslands – a systematic review. Basic Appl Ecol 88:9–18. https://doi.org/10.1016/j.baae.2025.08.003

Herzon I, Bullock JM, Humbert J-Y, Arlettaz R, Török P, Stein-Bachinger K (2026) Improving biodiversity in agricultural grassland systems. In: Advances in temperate grassland science and management. Burleigh Dodds Science Publishing, pp 341–382

Høye TT, Kresse J-C, Koltz AM, Bowden JJ (2020) Earlier springs enable high-Arctic wolf spiders to produce a second clutch. Proc R Soc B Biol Sci 287(1929):20200982. https://doi.org/10.1098/rspb.2020.0982

Humbert J-Y, Ghazoul J, Richner N, Walter T (2010) Hay harvesting causes high orthopteran mortality. Agric Ecosyst Environ 139(4):522–527. https://doi.org/10.1016/j.agee.2010.09.012

Hunter MD, Price PW (1992) Playing chutes and ladders: heterogeneity and the relative roles of bottom-up and top-down forces in natural communities. Ecology 73(3):724–732

Huntzinger M, Karban R, Cushman JH (2008) Negative effects of vertebrate herbivores on invertebrates in a coastal dune community. Ecology 89(7):1972–1980. https://doi.org/10.1890/07-0834.1

IPCC (2023) Climate Change 2023: Synthesis Report. Contribution of Working Groups I, II and III to the Sixth Assessment Report of the Intergovernmental Panel on

Climate Change [Core Writing Team, H. Lee and J. Romero (eds.)] (pp. 35–115). IPCC. https://doi.org/10.59327/IPCC/AR6-9789291691647

Isbell F, Reich PB, Tilman D, Hobbie SE, Polasky S, Binder S (2013) Nutrient enrichment, biodiversity loss, and consequent declines in ecosystem productivity. Proc Natl Acad Sci 110(29):11911–11916. https://doi.org/10.1073/pnas.1310880110

Jactel H, Brockerhoff EG (2007) Tree diversity reduces herbivory by forest insects. Ecol Lett 10(9):835–848. https://doi.org/10.1111/j.1461-0248.2007.01073.x

Jerrentrup JS, Wrage-Mönnig N, Röver K-U, Isselstein J (2014) Grazing intensity affects insect diversity via sward structure and heterogeneity in a long-term experiment. J Appl Ecol 51(4):968–977. https://doi.org/10.1111/1365-2664.12244

Jia S, Wang X, Yuan Z, Lin F, Ye J, Hao Z, Luskin MS (2018) Global signal of top-down control of terrestrial plant communities by herbivores. Proc Natl Acad Sci 115(24):6237. https://doi.org/10.1073/pnas.1707984115

Jiang H, Ke J, Zhang P, Dang Y, Liu X (2025) Top-down control of plant communities by consumers. J Ecol 113(10):3005–3018. https://doi.org/10.1111/1365-2745.70145

Johnson PTJ, Preston DL, Hoverman JT, Richgels KLD (2013) Biodiversity decreases disease through predictable changes in host community competence. Nature 494(7436):230–233. https://doi.org/10.1038/nature11883

Johnson SN, Cibils-Stewart X, Waterman JM, Biru FN, Rowe RC, Hartley SE (2022) Elevated atmospheric CO2 changes defence allocation in wheat but herbivore resistance persists. Proc R Soc B Biol Sci 289(1969):20212536. https://doi.org/10.1098/rspb.2021.2536

Kaspari M, Welti EAR (2024) Nutrient dilution and the future of herbivore populations. Trends Ecol Evol 39(9):809–820. https://doi.org/10.1016/j.tree.2024.05.001

Kempel A, Schädler M, Chrobock T, Fischer M, van Kleunen M (2011) Tradeoffs associated with constitutive and induced plant resistance against herbivory. Proc Natl Acad Sci 108(14):5685–5689. https://doi.org/10.1073/pnas.1016508108

Kempel A, Razanajatovo M, Stein C, Unsicker SB, Auge H, Weisser WW, Fischer M, Prati D (2015) Herbivore preference drives plant community composition. Ecology 96(11):2923–2934. https://doi.org/10.1890/14-2125.1

Kempel A, Allan E, Gossner MM, Jochum M, Grace JB, Wardle DA (2023) From bottom-up to top-down control of invertebrate herbivores in a retrogressive chronosequence. Ecol Lett 26(3):411–424. https://doi.org/10.1111/ele.14161

Kempel A, Adamidis GC, Anadón JD, Atkinson J, Auge H, Avtzis D, Bachelot B, Bashirzadeh M, Bota JL, Classen A, Constantinou I, Crawley M, de Bellis T, Dostal P, Ebeling A, Eisenhauer N, Eldridge DJ, Encina G, Estrada C et al (2025) The bug-network (BugNet): a global experimental network testing the effects of invertebrate herbivores and fungal pathogens on plant communities and ecosystem function in open ecosystems. Ecol Evol 15(10):e72111. https://doi.org/10.1002/ece3.72111

Kobori Y, Amano H (2003) Effect of rainfall on a population of the diamondback moth, Plutella xylostella (Lepidoptera: Plutellidae). Appl Entomol Zool 38(2):249–253. https://doi.org/10.1303/aez.2003.249

Koricheva J, Mulder CPH, Schmid B, Joshi J, Huss-Danell K (2000) Numerical responses of different trophic groups of invertebrates to manipulations of plant diversity in grasslands. Oecologia 125(2):271–282. https://doi.org/10.1007/s004420000450

Lantschner MV, Aukema BH, Corley JC (2019) Droughts drive outbreak dynamics of an invasive forest insect on an exotic host. For Ecol Manag 433:762–770. https://doi.org/10.1016/j.foreco.2018.11.044

Laws AN (2017) Climate change effects on predator–prey interactions. Curr Opin Insect Sci 23:28–34. https://doi.org/10.1016/j.cois.2017.06.010

Lawton JH (1983) Plant architecture and the diversity of phytophagous insects. Annu Rev Entomol 28:23–39. https://doi.org/10.1146/annurev.en.28.010183.000323

Lemoine NP, Drews WA, Burkepile DE, Parker JD (2013) Increased temperature alters feeding behavior of a generalist herbivore. Oikos 122(12):1669–1678. https://doi.org/10.1111/j.1600-0706.2013.00457.x

Lemoine NP, Burkepile DE, Parker JD (2014) Variable effects of temperature on insect herbivory. PeerJ 2:e376. https://doi.org/10.7717/peerj.376
Lemoine NP, Doublet D, Salminen J, Burkepile DE, Parker JD (2017) Responses of plant phenology, growth, defense, and reproduction to interactive effects of warming and insect herbivory. Ecology 98(7):1817–1828. https://doi.org/10.1002/ecy.1855
Liao C, Peng R, Luo Y, Zhou X, Wu X, Fang C, Chen J, Li B (2008) Altered ecosystem carbon and nitrogen cycles by plant invasion: a meta-analysis. New Phytol 177(3):706–714. https://doi.org/10.1111/j.1469-8137.2007.02290.x
Lind EM, Borer E, Seabloom E, Adler P, Bakker JD, Blumenthal DM, Crawley M, Davies K, Firn J, Gruner DS, Stanley Harpole W, Hautier Y, Hillebrand H, Knops J, Melbourne B, Mortensen B, Risch AC, Schuetz M, Stevens C, Wragg PD (2013) Life-history constraints in grassland plant species: a growth-defence trade-off is the norm. Ecol Lett 16(4):513–521. https://doi.org/10.1111/ele.12078
Lind EM, La Pierre KJ, Seabloom EW, Alberti J, Iribarne O, Firn J, Gruner DS, Kay AD, Pascal J, Wright JP, Yang L, Borer ET (2017) Increased grassland arthropod production with mammalian herbivory and eutrophication: a test of mediation pathways. Ecology 98(12):3022–3033. https://doi.org/10.1002/ecy.2029
Liu M, Jiang P, Chase JM, Liu X (2024) Global insect herbivory and its response to climate change. Curr Biol 34(12):2558–2569.e3. https://doi.org/10.1016/j.cub.2024.04.062
Mattson WJ (1980) Herbivory in relation to plant nitrogen content. Annu Rev Ecol Evol Syst 11:119–161. https://doi.org/10.1146/annurev.es.11.110180.001003
Mattson WJ, Haack RA (1987) The role of drought in outbreaks of plant-eating insects. Bioscience 37(2):110–118. https://doi.org/10.2307/1310365
Mensink PJ, Henry HAL (2011) Rain events influence short-term feeding preferences in the snail Cepaea nemoralis. J Molluscan Stud 77(3):241–247. https://doi.org/10.1093/mollus/eyr011
Milchunas DG, Sala OE, Lauenroth WK (1988) A generalized model of the effects of grazing by large herbivores on grassland community structure. Am Nat 132(1):87–106. https://doi.org/10.1086/284839
Mitchell C, Brennan RM, Graham J, Karley AJ (2016) Plant defense against herbivorous pests: exploiting resistance and tolerance traits for sustainable crop protection. Front Plant Sci 7:1132
Mordecai EA (2011) Pathogen impacts on plant communities: unifying theory, concepts, and empirical work. Ecol Monogr 81(3):429–441. https://doi.org/10.1890/10-2241.1
Moreira X, Abdala-Roberts L, Rasmann S, Castagneyrol B, Mooney KA (2016) Plant diversity effects on insect herbivores and their natural enemies: current thinking, recent findings, and future directions. Curr Opin Insect Sci 14:1–7. https://doi.org/10.1016/j.cois.2015.10.003
Murdoch WW, Evans FC, Peterson CH (1972) Diversity and pattern in plants and insects. Ecology 53(5):819–829. https://doi.org/10.2307/1934297
Oksanen L, Oksanen T (2000) The logic and realism of the hypothesis of exploitation ecosystems. Am Nat 155(6):703–723. https://doi.org/10.1086/303354
Oksanen T, Oksanen L, Vuorinen KEM, Wolf C, Mäkynen A, Olofsson J, Ripple WJ, Virtanen R, Utsi TA (2020) The impact of thermal seasonality on terrestrial endotherm food web dynamics: a revision of the exploitation ecosystem hypothesis. Ecography 43(12):1859–1877. https://doi.org/10.1111/ecog.05076
Olff H, Ritchie ME (1998) Effects of herbivores on grassland plant diversity. Trends Ecol Evol 13(7):261–265. https://doi.org/10.1016/s0169-5347(98)01364-0
Pacala SW, Crawley MJ (1992) Herbivores and plant diversity. Am Nat 140(2):243–260. https://doi.org/10.1086/285411
Panassiti B, Ewald J, Hofmann M, Trivellone V, Styrnik V, Nickel H, Neumayer J, Pospisil K, Klein D, Tobisch C, König S, Richter T, Geres LS, Baier R, Seibold S (2025) Effects of the timing of grazing on insect diversity and insect–plant interactions in mountain grasslands. Ecol Appl 35
Parmesan C (2006) Ecological and evolutionary responses to recent climate change. Annu Rev Ecol Evol Syst 37:637–669. https://doi.org/10.1146/annurev.ecolsys.37.091305.110100

Paseka RE, White LA, Van de Waal DB, Strauss AT, González AL, Everett RA, Peace A, Seabloom EW, Frenken T, Borer ET (2020) Disease-mediated ecosystem services: pathogens, plants, and people. Trends Ecol Evol 35(8):731–743. https://doi.org/10.1016/j.tree.2020.04.003

Petschenka G, Agrawal AA (2016) How herbivores coopt plant defenses: natural selection, specialization, and sequestration. Curr Opin Insect Sci 14:17–24. https://doi.org/10.1016/j.cois.2015.12.004

Pfisterer AB, Diemer M, Schmid B (2003) Dietary shift and lowered biomass gain of a generalist herbivore in species-poor experimental plant communities. Oecologia 135(2):234–241. https://doi.org/10.1007/s00442-002-1169-0

Pichon NA, Cappelli SL, Soliveres S, Hölzel N, Klaus VH, Kleinebecker T, Allan E (2020) Decomposition disentangled: a test of the multiple mechanisms by which nitrogen enrichment alters litter decomposition. Funct Ecol 34(7):1485–1496. https://doi.org/10.1111/1365-2435.13560

Post E, Pedersen C (2008) Opposing plant community responses to warming with and without herbivores. Proc Natl Acad Sci 105(34):12353–12358. https://doi.org/10.1073/pnas.0802421105

Powell KI, Chase JM, Knight TM (2011) A synthesis of plant invasion effects on biodiversity across spatial scales. Am J Bot 98(3):539–548. https://doi.org/10.3732/ajb.1000402

Price PW (1991) The plant vigor hypothesis and herbivore attack. Oikos 62(2):244–251. https://doi.org/10.2307/3545270

Pringle RM, Abraham JO, Anderson TM, Coverdale TC, Davies AB, Dutton CL, Gaylard A, Goheen JR, Holdo RM, Hutchinson MC, Kimuyu DM, Long RA, Subalusky AL, Veldhuis MP (2023) Impacts of large herbivores on terrestrial ecosystems. Curr Biol 33(11):R584–R610. https://doi.org/10.1016/j.cub.2023.04.024

Proske A, Lokatis S, Rolff J (2022) Impact of mowing frequency on arthropod abundance and diversity in urban habitats: a meta-analysis. Urban For Urban Green 76:127714. https://doi.org/10.1016/j.ufug.2022.127714

Quandahor P, Kim L, Kim M, Lee K, Kusi F, Jeong I (2024) Effects of agricultural pesticides on decline in insect species and individual numbers. Environments 11(8):182. https://doi.org/10.3390/environments11080182

Révész K, Gallé R, Humbert J-Y, Batáry P (2025) Effects of uncut refuge management on grassland arthropods – a systematic review. Global Ecol Conserv 57:e03381. https://doi.org/10.1016/j.gecco.2024.e03381

Robinson EA, Ryan GD, Newman JA (2012) A meta-analytical review of the effects of elevated CO2 on plant–arthropod interactions highlights the importance of interacting environmental and biological variables. New Phytol 194(2):321–336. https://doi.org/10.1111/j.1469-8137.2012.04074.x

Rogers A, Ainsworth EA, Leakey ADB (2009) Will elevated carbon dioxide concentration amplify the benefits of nitrogen fixation in legumes? Plant Physiol 151(3):1009–1016. https://doi.org/10.1104/pp.109.144113

Root RB (1973) Organization of a Plant-Arthropod Association in simple and diverse habitats: the Fauna of collards (brassica Oleracea). Ecol Monogr 43(1):95–124. https://doi.org/10.2307/1942161

Rosenzweig C, Iglesias A, Yang X et al (2001) Climate change and extreme weather events; implications for food production, plant diseases, and pests. Glob Chang Hum Health 2:90–104. https://doi.org/10.1023/A:1015086831467

Rottstock T, Joshi J, Kummer V, Fischer M (2014) Higher plant diversity promotes higher diversity of fungal pathogens, while it decreases pathogen infection per plant. Ecology 95(7):1907–1917. https://doi.org/10.1890/13-2317.1

Roy BA, Güsewell S, Harte J (2004) Response of plant pathogens and herbivores to a warming experiment. Ecology 85(9):2570–2581. https://doi.org/10.1890/03-0182

Sardans J, Rivas-Ubach A, Peñuelas J (2012) The C:N:P stoichiometry of organisms and ecosystems in a changing world: a review and perspectives. Perspect Plant Ecol Evol Syst 14(1):33–47. https://doi.org/10.1016/j.ppees.2011.08.002

Schädler M, Jung G, Auge H, Brandl R (2003) Does the Fretwell–Oksanen model apply to invertebrates? Oikos 100(1):203–207. https://doi.org/10.1034/j.1600-0706.2003.11821.x

Scherber C, Eisenhauer N, Weisser WW, Schmid B, Voigt W, Fischer M, Schulze E-D, Roscher C, Weigelt A, Allan E, Beßler H, Bonkowski M, Buchmann N, Buscot F, Clement LW, Ebeling A, Engels C, Halle S, Kertscher I et al (2010) Bottom-up effects of plant diversity on multitrophic interactions in a biodiversity experiment. Nature 468(7323):553–556. https://doi.org/10.1038/nature09492

Seabloom EW, Kinkel L, Borer ET, Hautier Y, Montgomery RA, Tilman D (2017) Food webs obscure the strength of plant diversity effects on primary productivity. Ecol Lett 20(4):505–512. https://doi.org/10.1111/ele.12754

Skinner GLV, Cooke R, Roy HE, Isaac NJB, Outhwaite CL, Rodger J, Millard J (2026) Meta-analysis reveals negative but highly variable impacts of invasive alien species across terrestrial insect orders. Nat Commun 17(1). https://doi.org/10.1038/s41467-025-67925-9

Smith MD, Wilkins KD, Holdrege MC, Wilfahrt P, Collins SL, Knapp AK, Sala OE, Dukes JS, Phillips RP, Yahdjian L, Gherardi LA, Ohlert T, Beier C, Fraser LH, Jentsch A, Loik ME, Maestre FT, Power SA, Yu Q et al (2024) Extreme drought impacts have been underestimated in grasslands and shrublands globally. Proc Natl Acad Sci 121(4):e2309881120. https://doi.org/10.1073/pnas.2309881120

Southwood TRE, Brown VK, Reader PM (1979) The relationships of plant and insect diversities in succession. Biol J Linn Soc 12(4):327–348. https://doi.org/10.1111/j.1095-8312.1979.tb00063.x

Srivastava DS, Lawton JH (1998) Why more productive sites have more species: an experimental test of theory using tree-hole communities. Am Nat 152(4):510–529. https://doi.org/10.1086/286187

Stamp N (2003) Out of the quagmire of plant defense hypotheses. Q Rev Biol 78(1):23–55. https://doi.org/10.1086/367580

Steidle JLM, Kimmich T, Csader M, Betz O (2022) Negative impact of roadside mowing on arthropod fauna and its reduction with ‘arthropod-friendly’ mowing technique. J Appl Entomol 146(5):465–472. https://doi.org/10.1111/jen.12976

Stevens CJ, Dise NB, Mountford JO, Gowing DJ (2004) Impact of nitrogen deposition on the species richness of grasslands. Science 303(5665):1876–1879. https://doi.org/10.1126/science.1094678

Stiling P, Cornelissen T (2007) How does elevated carbon dioxide (CO2) affect plant–herbivore interactions? A field experiment and meta-analysis of CO2-mediated changes on plant chemistry and herbivore performance. Glob Chang Biol 13(9):1823–1842. https://doi.org/10.1111/j.1365-2486.2007.01392.x

Sun S-J, Lee X-Y, Wang Y-J, Chuang W-P, Hwang S-Y, Ho C-K (2025) Understanding crop performance and pest control under climate change requires considering interactions among warming, elevated CO2, and trophic interactions. Sci Rep 15(1):39519. https://doi.org/10.1038/s41598-025-23639-y

Suominen O, Danell K, Bergstrom R (1999) Moose, trees, and ground-living invertebrates: indirect interactions in Swedish pine forests. Oikos 84(2):215. https://doi.org/10.2307/3546716

Suttle KB, Thomsen MA, Power ME (2007) Species interactions reverse grassland responses to changing climate. Science 315(5812):640–642. https://doi.org/10.1126/science.1136401

Tallamy DW, Shropshire KJ (2009) Ranking lepidopteran use of native versus introduced plants. Conserv Biol 23(4):941–947. https://doi.org/10.1111/j.1523-1739.2009.01202.x

Thakur MP, Reich PB, Hobbie SE, Stefanski A, Rich R, Rice KE, Eddy WC, Eisenhauer N (2018) Reduced feeding activity of soil detritivores under warmer and drier conditions. Nat Clim Chang 8(1):75–78. https://doi.org/10.1038/s41558-017-0032-6

Thakur MP, Gu Z, van Kleunen M, Zhou X (2025) Invasion impacts in terrestrial ecosystems: global patterns and predictors. Science 390(6771):381–385. https://doi.org/10.1126/science.adq3101

Throop HL, Lerdau MT (2004) Effects of nitrogen deposition on insect herbivory: implications for community and ecosystem processes. Ecosystems 7(2). https://doi.org/10.1007/s10021-003-0225-x

Tylianakis JM, Didham RK, Bascompte J, Wardle DA (2008) Global change and species interactions in terrestrial ecosystems. Ecol Lett 11(12):1351–1363. https://doi.org/10.1111/j.1461-0248.2008.01250.x

van Klink R, van der Plas F, van Noordwijk CGE et al (2014) Effects of large herbivores on grassland arthropod diversity. Biol Rev 90(2):347–366. https://doi.org/10.1111/brv.12113

Vogel A, Scherer-Lorenzen M, Weigelt A (2012) Grassland resistance and resilience after drought depends on management intensity and species richness. PLoS One 7(5):e36992. https://doi.org/10.1371/journal.pone.0036992

Wan N-F, Fu L, Dainese M, Kiær LP, Hu Y-Q, Xin F, Goulson D, Woodcock BA, Vanbergen AJ, Spurgeon DJ, Shen S, Scherber C (2025) Pesticides have negative effects on non-target organisms. Nat Commun 16(1):1360. https://doi.org/10.1038/s41467-025-56732-x

Weigelt A, Weisser WW, Buchmann N, Scherer-Lorenzen M (2009) Biodiversity for multifunctional grasslands: equal productivity in high-diversity low-input and low-diversity high-input systems. Biogeosciences 6(8):1695–1706. https://doi.org/10.5194/bg-6-1695-2009

Welti EAR, Sanders NJ, de Beurs KM, Kaspari M (2019) A distributed experiment demonstrates widespread sodium limitation in grassland food webs. Ecology 100(3):e02600. https://doi.org/10.1002/ecy.2600

Welti EAR, Roeder KA, de Beurs KM, Joern A, Kaspari M (2020) Nutrient dilution and climate cycles underlie declines in a dominant insect herbivore. Proc Natl Acad Sci 117(13):7271–7275. https://doi.org/10.1073/pnas.1920012117

White TCR (1993) The Inadequate Environment: Nitrogen and the Abundance of Animals. https://api.semanticscholar.org/CorpusID:82166633

Woodcock BA, Isaac NJB, Bullock JM, Roy DB, Garthwaite DG, Crowe A, Pywell RF (2016) Impacts of neonicotinoid use on long-term population changes in wild bees in England. Nat Commun 7(1):12459. https://doi.org/10.1038/ncomms12459

Wright IJ, Reich PB, Westoby M, Ackerly DD, Baruch Z, Bongers F, Cavender-Bares J, Chapin T, Cornelissen JHC, Diemer M, Flexas J, Garnier E, Groom PK, Gulias J, Hikosaka K, Lamont BB, Lee T, Lee W, Lusk C et al (2004) The worldwide leaf economics spectrum. Nature 428(6985):821–827. https://doi.org/10.1038/nature02403

Yan P, He N, Fernández-Martínez M, Yang X, Zuo Y, Zhang H, Wang J, Chen S, Song J, Li G, Valencia E, Wan S, Jiang L (2025) Plant acquisitive strategies promote resistance and temporal stability of semiarid grasslands. Ecol Lett 28(4):e70110. https://doi.org/10.1111/ele.70110

Joint Effects of Tree Diversity and Drought on Insect Herbivory: Insights from a Long-Term Tree Diversity Experiment

6

Bastien Castagneyrol, Tom Barlier, Tony Chahine, Andrea Galmán, Noah La Paglia, Carla Vázquez-González, Charlotte Poeydebat, Thomas Ribot, Inge van Halder, Margaux Walbrecq, and Hervé Jactel

Abstract

Drought has long been recognised as a key abiotic driver of tree–herbivore interactions because of its effects on plant physiology, growth, and defensive traits. At the same time, tree diversity shapes forest water dynamics by altering resource uptake, hydraulic complementarity, and belowground competition while also influencing patterns of insect herbivory through associational processes. Consequently, drought is expected not only to directly affect tree–herbivore interactions by shaping plant traits at the level of individual trees but also to modulate how tree stands—via tree diversity—influence herbivore damage. Despite this conceptual expectation, empirical tests remain limited. This gap largely reflects the logistical challenges of experimentally manipulating water

B. Castagneyrol (✉) · T. Barlier · N. La Paglia · T. Ribot · I. van Halder · M. Walbrecq · H. Jactel
BIOGECO, INRAE, University of Bordeaux, Cestas, France
e-mail: bastien.castagneyrol@inrae.fr

T. Chahine
Department of Agricultural Sciences, University of Sassari, Sassari, Italy

A. Galmán
Department of Land Resources & Environmental Sciences, Montana State University, Bozeman, MT, USA

C. Vázquez-González
Misión Biológica de Galicia (MBG-CSIC), Pontevedra, Galicia, Spain

C. Poeydebat
Bordeaux Science Agro, INRAE, SAVE, ISVV, Gradignan, France

X. Moreira et al. (eds.), *Effects of Plant Diversity on Plant-Herbivore Interactions*, Ecological Studies 255, https://doi.org/10.1007/978-3-032-26461-9_6

availability in forest systems and the scarcity of long-term datasets that allow repeated assessments of diversity effects under contrasting hydrological conditions. Here, we synthesise insights from a decade-long tree diversity experiment in south-western France that incorporates controlled irrigation treatments. Focusing on and birch and three oak species growing in monocultures, mixed broadleaved stands, and mixtures of broadleaved species with pines, we evaluate how water availability interacts with tree diversity and species composition to shape insect herbivory. Our results reveal strong species-specific responses to drought, as well as species-specific patterns of associational susceptibility (with resistance being less common) driven primarily by tree species composition rather than diversity per se, and further modulated by drought. These findings challenge generalised predictions about tree diversity-mediated resistance to abiotic disturbances and underscore the importance of identifying the underlying mechanisms behind context-dependent outcomes in drought–diversity–herbivory relationships in forest ecosystems.

6.1 Introduction

As early as the nineteenth century, forestry manuals recommended mixing tree species to reduce the spread of pests and diseases (Boppe 1889). What began as empirical silvicultural knowledge has since been supported by a large body of observational studies (Plath et al. 2011; Guyot et al. 2016), experiments (Haase et al. 2015; Poeydebat et al. 2021), modelling approaches, and quantitative syntheses (Leonard et al. 2023). Across experimental plantations and mature forests alike, a recurrent pattern has emerged: Trees growing in heterospecific neighbourhoods tend, on average, to sustain lower insect loads and suffer less herbivore damage than trees in monocultures (Jactel et al. 2021). This pattern, termed associational resistance, has become a cornerstone of research on biodiversity–ecosystem functioning relationships in forests. However, alongside associational resistance, the opposite outcome—associational susceptibility—has also been documented (Schuldt et al. 2010; Wein et al. 2016). Although less frequent, associational susceptibility demonstrates that tree diversity does not universally protect against herbivory. Several sources of heterogeneity among studies have been explored, including tree species composition (Setiawan et al. 2014), herbivore identity and feeding guild (Jia et al. 2024), herbivore population density (Bognounou et al. 2017), and large-scale climatic context (Kambach et al. 2016). Accordingly, predicting when diversity will reduce or increase herbivory remains a major scientific challenge, particularly under rapid environmental change.

Three main, non-exclusive mechanisms have been proposed to explain associational effects. The first involves reduced host accessibility. In mixed stands, the absolute and relative abundance of host and non-host species determines both resource concentration and the probability of host detection (Castagneyrol et al. 2013, 2014, 2020; Klapwijk and Bjorkman 2018). Associational resistance is more likely when host trees are diluted among non-host neighbours, particularly when those neighbours are phylogenetically distant (Castagneyrol et al. 2014; de Groot et al. 2023). By contrast, mixtures containing closely related or alternative host species may enhance resource availability and increase herbivore pressure. The second mechanism concerns enhanced top-down control. Greater tree diversity may promote richer and more abundant communities of predators and parasitoids, thereby strengthening top-down regulation of herbivores (Staab and Schuldt 2020; Stemmelen et al. 2022). Increased structural complexity, diversified prey resources, and buffered microclimates in mixed stands are non-exclusive mechanisms favouring these natural enemies of herbivores. Yet empirical support remains equivocal, with some studies reporting stronger enemy effects in diverse stands and others finding weak or inconsistent responses (Stemmelen et al. 2022). The third mechanism centres on plant trait-mediated processes. Herbivory results from the consumption of plant tissues, but plant traits governing palatability and defence are dynamic (Castagneyrol et al. 2018b). Herbivore attack induces direct defences, such as the production of secondary metabolites (Karban 2011), and indirect defences, including the emission of volatile organic compounds (VOCs) that attract natural enemies (Rasheed et al. 2023). At the same time, neighbour identity, phylogenetic relatedness, and spatial proximity influence plant trait expression, all of these features which have been shown to be influenced by tree diversity (Rosado-Sánchez et al. 2017; Muiruri et al. 2019; Damestoy et al. 2021; Wang et al. 2024). Through these plant trait-mediated indirect pathways, tree diversity can reshape plant–insect interactions independently of changes in host plant density driving resource concentration effects.

Crucially, the above mechanisms further operate within an abiotic context that directly and indirectly shapes trophic interactions (Holopainen et al. 2018; Jactel et al. 2019; Vazquez-Gonzalez et al. 2024). Among abiotic drivers, drought plays a particularly prominent role. Water stress modifies plant physiology, often leading to the accumulation of secondary metabolites and free amino acids (Mundim and Pringle 2018; Lin et al. 2023). These changes can have contrasting consequences depending on the herbivore feeding guild. Many defoliators perform less well on severely water-stressed trees, whereas moderate drought may favour sap-sucking insects and leaf miners (Jactel et al. 2012; Bansal 2015; Gely et al. 2020). Drought and heat also alter VOC emission rates and atmospheric diffusion, potentially influencing natural enemy attraction and top-down control (Vázquez-González et al. 2022). In turn, tree diversity influences forest water dynamics. Mixed stands often exhibit enhanced fine-root development and partial niche partitioning, improving access to water under limiting conditions (Grossiord 2020; Jing et al. 2021; Zeng et al. 2021). Plant-plant heterospecific interactions may further induce adjustments in sap flow and leaf metabolism, while diverse canopies buffer microclimatic extremes (Schnabel et al. 2025). Accordingly, by mitigating or modifying plant water stress, tree diversity would be expected to interact with drought in shaping herbivore performance and damage. Nonetheless, empirical evidence for such interactive effects remains limited due to the paucity of experimental studies. Detecting them requires long-term datasets spanning multiple drought events or replicated experimental designs along climatic gradients—approaches that are often logistically and financially demanding. Alternatively, manipulating both tree diversity and water availability within a single experimental platform while also resource intensive could be a logistically easier and cost-effective approach to shed mechanistic insight based on local-scale studies.

In this chapter, we synthesise results from a decade-long study conducted in the ORPHEE tree diversity experiment in south-western France, where half of the plots have been irrigated while the other half remain under ambient rainfall. This design enables the assessment of both independent and interactive effects of tree species diversity and water availability on insect herbivory. We first describe the experimental framework and monitoring protocols. We then present two case studies that integrate published and unpublished data to extend the temporal scope of previous analyses. Rather than confirming a universal abiotic buffering role of diversity, our findings highlight strong context-dependency in tree diversity–drought interactions. Understanding these dynamics is critical in the context of climate change. As drought events intensify in frequency and severity, predicting forest vulnerability will require studying biotic and abiotic drivers. Long-term manipulations of both diversity and water supply can contribute critical knowledge to this end while also providing a realistic assessment of how associational effects emerge, persist, or reverse under changing environmental conditions.

6.2 The ORPHEE Experiment

The ORPHEE tree diversity experiment was established in 2008 in south-western France (44°44′ N, 00°46′ W) on a former maritime pine plantation growing on a sandy podzol (Castagneyrol et al. 2013). Prior to planting, pine stumps were removed, and the site was ploughed and fertilised with phosphorus and potassium to homogenise soil conditions. The experimental design comprises 8 replicated blocks, each containing 32 plots of 100 trees arranged in a 10 × 10 grid, with 2 m spacing between trees within each plot. Plots are separated by 3 m and randomly distributed within each block. The experiment follows a substitutive design manipulating tree species richness and composition using five native species: silver birch (*Betula pendula*), pedunculate oak (*Quercus robur*), Pyrenean oak (*Q. pyrenaica*), holm oak (*Q. ilex*), and maritime pine (*Pinus pinaster*). Plots contain 1 to 5 species, and all 31 possible species combinations were established, with the 5-species mixture replicated twice per block. Within each plot, trees are arranged in a regular and equiproportional pattern, so that individuals of a given composition experience identical neighbourhood structure and species proportions. In total, the experiment included 256 plots (8 blocks × 32 plots), initially encompassing 25,600 trees over 12 hectares.

The site was fenced to exclude large mammalian herbivores. Trees that died during the first growing season were replaced the following year to maintain design integrity and initial planting densities. During the early years, herbaceous vegetation was mown annually both within and between plots. Later, mowing was restricted to inter-plot areas to facilitate access, allowing understory vegetation to develop naturally within plots. An exception occurred in 2017, when European gorse (*Ulex europaeus*) and dwarf gorse (*U. minor*)—which had colonised much of the site—were manually removed to restore access to the trees.

In 2015, a large-scale irrigation treatment was implemented to test interactions between tree diversity and water availability. Four of the eight blocks were randomly assigned to nightly irrigation from May to October. Approximately 42 m^3 of water per plot were applied each night using 2.5 m-high positioned sprinklers, corresponding to roughly 3 mm of additional precipitation. Irrigation volumes were calculated based on regional climatic and soil data to alleviate summer water deficits without creating waterlogging. The effectiveness of the treatment was verified by measuring predawn leaf water potential in August 2015 and 2017 using a Scholander-type pressure chamber, confirming significant differences in water stress between irrigated and control plots.

6.2.1 Case Studies

Previous research in other systems has shown that the effects of tree diversity on insect herbivores vary among species depending on their diet breadth and feeding guild, two biological attributes that are partly correlated (Novotny et al. 2010). In particular, studies have shown that specialist and generalist herbivores, as well as defoliators, sap-feeders, and miners, can respond differently to host dilution and neighbourhood composition (Vehviläinen et al. 2007; Castagneyrol et al. 2014; Guyot et al. 2019; Jia et al. 2024). To investigate how tree diversity interacts with drought to shape herbivory, via changes in associational or plant trait-mediated mechanisms, we adopted complementary approaches targeting different types of herbivore damage.

In many cases, leaf damage could not be assigned to a particular insect species. We therefore quantified the percentage of leaf area removed as an integrative measure of defoliation (Case study 1: Defoliators). Although insects were not systematically sampled, field observations indicated that most damage was caused by generalist caterpillars and sawfly larvae and more occasionally by weevils and other Coleoptera. Other pests in the study area, however, are easily identifiable. For example, the invasive oak lace bug (*Corythucha arcuata*) causes characteristic symptoms on pedunculate and Pyrenean oaks. In this case, we quantified infestation by recording the number of attacked trees (Case study 2: Oak lace bug). Sampling details are summarised in Fig. 6.1.

All analyses were conducted in R (R Core Team 2024) using functions provided by the packages *glmmTMB* (Brooks et al. 2017) and *DHARMa* (Hartig 2016). For each case study, we applied the same modelling strategy. We used generalised linear mixed-effects models, adapting the error distribution and random effects to the response variable and the hierarchical structure of the data (Table 6.1).

We analysed the data within an information-theoretic framework, comparing candidate models based on their relative support. In all models, sampling year was included as a random factor to account for interannual variation in insect population

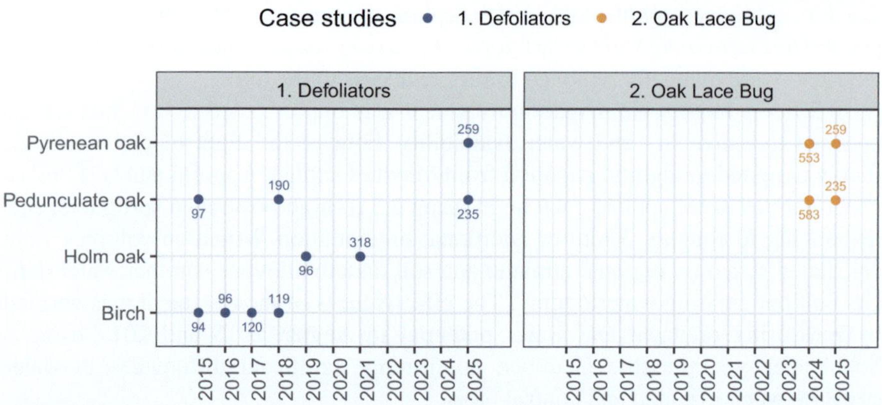

Fig. 6.1 Summary of the sampling design indicating which herbivore damage or pest presence was recorded each year for each species. Dots indicate that data were acquired. Numbers below and above the dots indicate the number of individual data points collected

Table 6.1 Summary of the characteristics of the GLMMs used to analyse data for each of the two case studies. Year is treated as a factor. Plot_ID refers to unique ID per plot, the coding of which ensuring that Plot_ID was nested within blocks

Case study	Response variable	Error distribution	Random effects
1	% leaf area (log-transformed)	Gaussian	(Year) + (Block:Irrigation) + (Block/Plot_ID)
2	Number of attacked vs. non-attacked trees	Binomial	(Year) + (Block:Irrigation) + (Block/Plot_ID)

densities and potential climatic anomalies. To reflect the hierarchical sampling design and repeated measurements within plots, we specified block and plot nested within block as random effects (Block/Plot_ID in R syntax; Table 6.1). Because reliable information on tree identity was unavailable, we assumed repeated measurements were independent within plots. As irrigation was applied at the block level, we incorporated the interaction between block and irrigation (Block:Irrigation; Table 6.1) as an additional random term to properly model the split-plot design.

We first fitted a beyond-optimal (global) model including species identity, irrigation, tree species richness, and tree species composition (a three-level factor: monocultures, mixtures without pines, and mixtures with pines), as well as all two- and three-way interactions among these predictors. We then generated all models nested within this global model and compared them using Akaike's Information Criterion corrected for small sample size (AICc; Burnham and Anderson 2002). Because our primary goal was to contrast the relative importance of tree species richness versus tree species composition in explaining insect herbivory, we excluded models that contained both variables simultaneously. Models within two AICc units of the best-supported model (i.e. the most parsimonious model with the lowest AICc) were considered equally plausible. When multiple models fell within this range (ΔAICc

<2), we performed model averaging, weighting coefficients by their AICc weights (w_i). We report averaged coefficients and variable importance, calculated as the sum of weights (SW), which reflects the probability that a given predictor occurs in the best-supported model.

6.2.1.1 Case Study 1: Defoliators

The same sampling protocol and scoring procedures were applied to birches following previous studies (e.g. Castagneyrol et al. 2018a, b; Poeydebat et al. 2021) and to oaks (Castagneyrol et al. 2017; Field et al. 2020; Galmán et al. 2022). However, the number of leaves assessed per tree was adjusted according to tree age and height to ensure comparable sampling effort among individuals. We visually estimated leaf area loss caused by chewing and skeletonising insect herbivores on randomly selected trees and leaves. Leaf-level herbivory scores were then averaged to obtain a single estimate per tree. Overall, leaf area loss fell within the range typically considered background herbivory (Kozlov and Zvereva 2017). Across years, species, and individuals, damage generally ranged between 0 and 10% (Fig. 6.4). On average, herbivory was approximately twice as high in birch and Pyrenean oak compared to holm oak and about five times higher in pedunculate oak (Fig. 6.2).

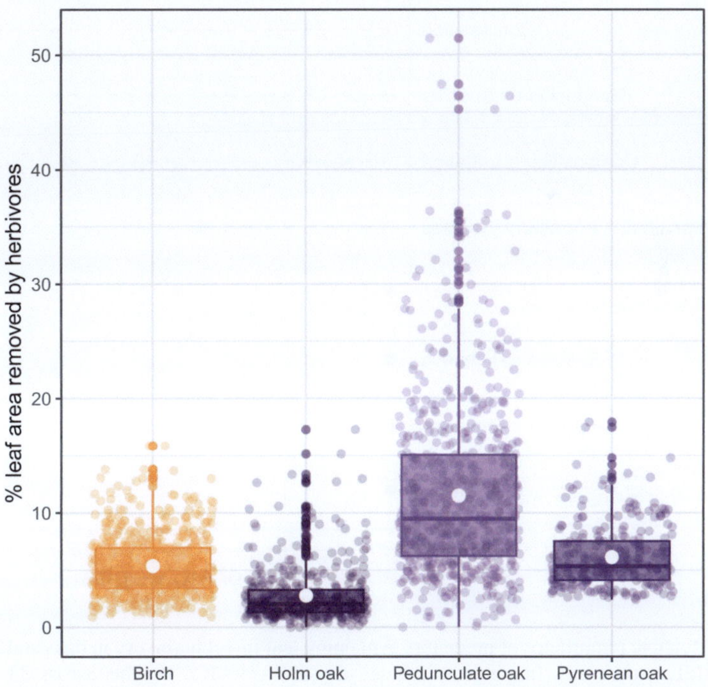

Fig. 6.2 Mean percentage of leaf area removed by insect herbivores for each surveyed species. Light, transparent dots represent raw data. White dots indicate the overall mean. Boxplots show the first and third quartiles, the median (second quartile), and extreme values (whiskers)

We compared a set of 33 candidate models using ΔAICc to evaluate their relative support. Model selection revealed strong support for three competing models (ΔAICc <2). These top-ranked models consistently included tree species composition, species identity, and irrigation, as well as all two-way interactions among these predictors and the three-way interaction. Each of these main effects and two-way interactions showed high relative importance ($w_i > 0.75$), indicating that they jointly structure variation in the response variable. In contrast, tree species richness—either as a main effect or in interaction with other predictors—received very low support ($w_i < 0.05$), suggesting a negligible contribution to explaining insect herbivory (Fig. 6.3).

The effect of tree species diversity on herbivory was generally weak and context dependent, whereas tree species identity and plot species composition effects were

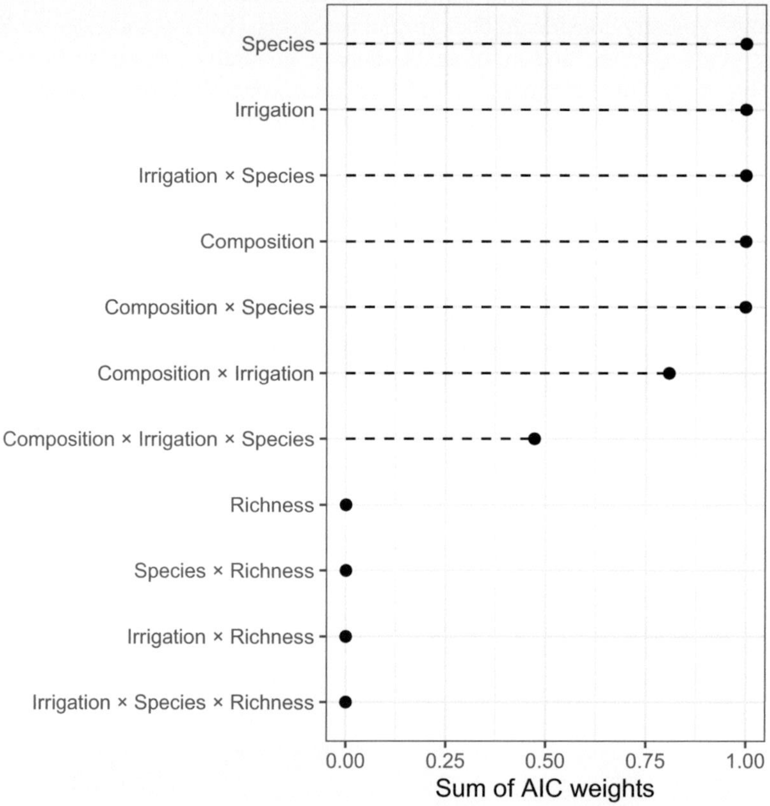

Fig. 6.3 Relative importance of predictors explaining leaf insect herbivory in oaks and birches in the ORPHEE experiment. Importance is based on summed AICc weights across 33 candidate models. Shown are species identity (Species), tree species richness (Richness), tree species composition (Composition), irrigation, and their two- and three-way interactions. Higher summed weights indicate stronger overall support for a given predictor in explaining variation in the response variable

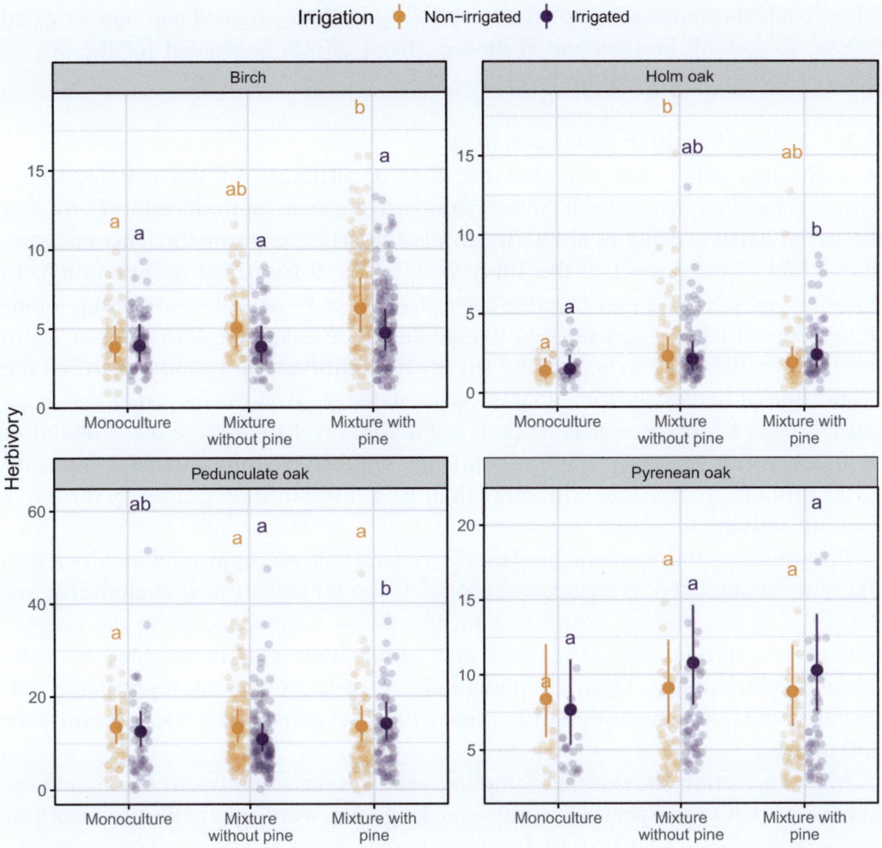

Fig. 6.4 Effects of tree species diversity and composition (the presence and absence of pines in mixtures) and irrigation on leaf insect herbivory in four broadleaved species in the ORPHEE experiment. Semi-transparent dots represent raw data. Dark dots and error bars show predictions from the best-supported model identified through model selection. Letters above the bars indicate contrasts among composition treatments, estimated separately for each species and irrigation level. Different letters denote significant differences between treatments. Note that the vertical scale differs among species

strong but contingent on irrigation (Fig. 6.4). The effect of irrigation varied among tree species. For birch, herbivory tended to be higher in non-irrigated (presumably more stressed) trees, although the difference was small (<2% leaf area loss) and statistically significant only in mixtures with pines. A similar trend was observed in pedunculate oak, but significance was detected only in mixtures without pines. In contrast, holm oak exhibited slightly higher herbivory in irrigated blocks, with a small effect size and significance restricted to mixtures with pines. In addition, patterns suggestive of associational susceptibility were contingent on irrigation; in particular, the presence of pines correlated with increased insect herbivory in non-irrigated birches and in irrigated pedunculate oaks (Fig. 6.4). Holm oak, on the other hand, showed associational susceptibility in both irrigated and non-irrigated blocks. Across all best-supported models, fixed effects accounted for the largest share of variation in insect herbivory.

6.2.1.2 Case Study 2: Oak Lace Bug

In 2024 and 2025, we recorded the first occurrences of the oak lace bug, *Corythucha arcuata*, a North American invasive species introduced to Europe in the early 2010s (Csóka et al. 2020), in the ORPHEE experiment. This oak specialist had already reached the study area by 2019 (personal observations). In 2024, we assessed its presence/absence in summer by visually inspecting a random sample of five Pyrenean and five pedunculate oaks per plot for up to 2 min per tree. In 2025, we extended the survey to additional plots and quantified the proportion of leaves showing characteristic damage. We first converted leaf-level observations to the presence/absence at the tree level to analyse the probability of attack, using year as a temporal replicate. We then examined damage intensity as the proportion of leaves with more than 10% discoloration caused by the bug's sucking activity.

Preliminary analyses indicated that Pyrenean oak was approximately twice as likely to be attacked as pedunculate oak. To better reflect host availability, we defined a new variable—resource concentration—corresponding to the proportion of susceptible trees (Pyrenean and pedunculate oaks), weighted by their relative vulnerability. Thus, a monoculture of pedunculate oak represented 100 resource units, a monoculture of Pyrenean oak 200 units, and a 2-species mixture 150 units.

Model selection showed that sampling year, species identity, irrigation, and the interaction between species identity and irrigation were important predictors of attack probability (ΔAICc >0.9). In contrast, tree species richness and resource concentration had negligible influences (all $w_i < 0.40$). In particular, we found that the probability of attack was consistently higher in non-irrigated than irrigated plots (particularly for Pedunculate oak) and about twice as high in Pyrenean compared to pedunculate oak (Fig. 6.5). Results were qualitatively similar when analysing damage intensity at the tree level (not shown).

6.3 Discussion

Our 10-year analysis of insect herbivory in the ORPHEE tree diversity experiment with controlling for water supply revealed, contrary to expectations, that the effects of tree diversity and drought were weak, inconsistent, and in some cases counterintuitive. Results, particularly for irrigation, were also somewhat contingent on the herbivore species or guild. Indeed, on average, per cent of leaf area consumed by defoliators was statistically indistinguishable between monocultures and mixed plots; inconsistent because responses varied markedly among species; and contradictory because, contrary to predicted associational resistance, we observed a trend for associational susceptibility, albeit contingent on the tree species and drought. In

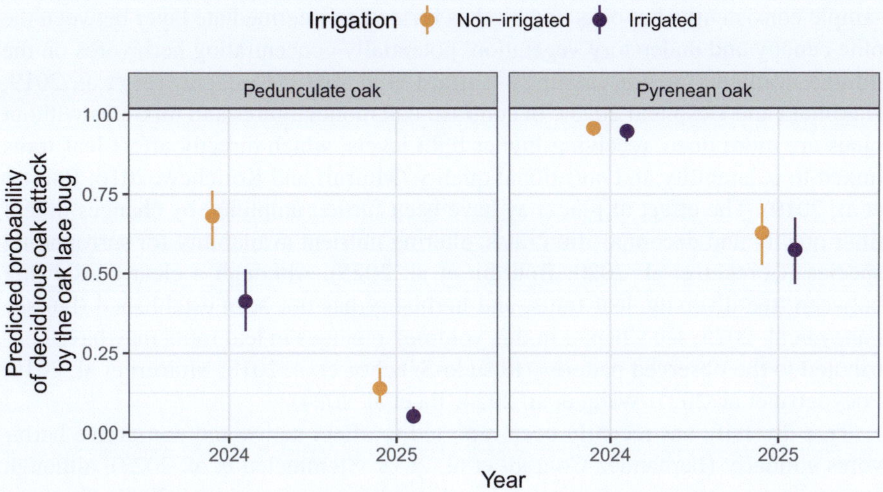

Fig. 6.5 Probability of attack by the oak lace bug on pedunculate and Pyrenean oaks in irrigated and non-irrigated plots

contrast, the oak lace bug, a specialist piercing-sucking herbivore, partly followed expected patterns: Its probability of attack increased under water stress, yet tree diversity and species composition again had no detectable effect.

A number of studies have found that the composition of tree assemblages exerts a stronger influence on insect herbivory compared to tree species richness (Schuldt and Scherer-Lorenzen 2014; Hertzog et al. 2018). Previous studies in the ORPHEE experiment highlighted the presence of pine as a key driver of defoliation by chewing insects in mixed tree species plots, a pattern reinforced by both published and unpublished datasets. However, the associational susceptibility caused by pines was unexpected. The ORPHEE experiment, established in 2009 with even-aged, 1-year-old seedlings, implemented the irrigation treatment from 2015 for 10 years. During the early survey period, trees were still young (7 years old). Fast-growing species such as silver birch and maritime pine rapidly developed canopies that overtopped slower-growing oaks, and by 2014, pines even exceeded birches in height (Castagneyrol et al. 2020). These dynamics have important implications for interpreting the observed herbivory patterns.

First, the presence of pine likely reduced the apparency of other trees in mixed plots throughout the survey period by creating physical barriers that limited colonisation of smaller trees by insect herbivores (Castagneyrol et al. 2013, 2020; Yuan and Wang 2025). Oak apparency was further diminished by the presence of birch (Castagneyrol et al. 2017). Therefore, the gradual reduction in birch and oak apparency as trees grew is unlikely to explain the observed herbivory patterns, which would have predicted increasing associational resistance over time. Instead, the vertical structure of the plots may have influenced herbivory and tree diversity outcomes. Despite slow growth, most oaks eventually overtopped the understory. Some oaks remain very small (<50 cm, 17 years post-planting) and were difficult to sample consistently, but oaks and birches formed an intermediate layer between the pine canopy and understory vegetation, potentially concentrating herbivores on the limited available trees in mixtures (Giffard et al. 2012; Castagneyrol et al. 2019; Hakimara and Despland 2025). In contrast, oak monocultures and mixtures without pines are more open, receiving higher light levels, which directly affect leaf traits linked to palatability and nutritional quality (Muiruri and Koricheva 2016; Poorter et al. 2019). The effect of pine may have been further amplified by changes in leaf litter quality and decomposition rates, altering nutrient availability for surrounding species (Corcket et al. 2020; Bourdin et al. 2025). Although a clear causal link between tree diversity, leaf traits, and herbivory has not been established (but see Fahey et al. 2025; see Chap. 2 in this volume), changes in leaf traits may have contributed to the observed patterns (Rosado-Sánchez et al. 2018; Muiruri et al. 2019; Poeydebat et al. 2021; Wang et al. 2024; Jia et al. 2024).

Tree diversity has recently been reported to affect native and non-native herbivores similarly (Fernandez-Conradi et al. 2018; Stemmelen et al. 2023), although associational outcomes may not necessarily be the same (Jactel et al. 2025). Contrary to expectations (Jactel et al. 2021; Hoch et al. 2024), however, we found no detectable effect of tree diversity on the probability of oak attack by the oak lace bug. This insect, a North American specialist herbivore (Paulin et al. 2020), was introduced to south-western France in 2018 and first observed in the ORPHEE experiment in 2023, prompting surveys the following year. During early establishment phase, still low population density might have generated stochastic processes override mechanistic effects, potentially obscuring any influence of tree diversity or irrigation. This is compounded by the fact that this insect exhibits limited dispersal, which may limit the detection of diversity effects. However, the fact that oak lace bug occurrence was higher on non-irrigated trees suggests a preference for water-stressed hosts and that dispersal constraints were not particularly strong to prevent this result reflecting better performance and increased attack rates by this insect on irrigated plots. Future targeted experiments will be required to resolve this question. The sandy soils of the ORPHEE experiment create highly variable water availability depending on natural precipitation. Irrigated plots received a consistent nightly water supply during the growing season to mitigate drought effects. Although soil moisture was not directly monitored, non-irrigated trees likely experienced intermittent stress, alternating between severe drought and relief after rain, while irrigated trees, if stressed, experienced relatively constant moderate stress. Although theory provides limited guidance on how different drought regimes affect chewing insects (Gely et al. 2020), we expected that these differences would translate into observable differences in herbivory. Surprisingly, herbivory was neither lower nor less variable in irrigated plots, where water supply was more continuous.

We can only speculate about the causes of the weak effect of irrigation on insect herbivory. First, plant responses to drought involve a range of physiological and chemical changes in leaves that vary even among closely related species (Xiao et al. 2024) and can have opposing effects on herbivores (Gutbrodt et al. 2011; Sahu and Giri 2025). Reduced leaf turgor may limit nitrogen-containing compounds essential for herbivores while increasing exposure to chemical defences (Mundim and Pringle

2018). Conversely, osmotic adjustments—such as the accumulation of free amino acids and soluble sugars—can enhance leaf nutritional quality (Gely et al. 2020). Offsetting effects of these mechanisms may therefore result in weak net differences in herbivory. Second, irrigation strongly influenced understory vegetation, which was denser and more abundant in irrigated plots (Castagneyrol et al. 2017; Corcket et al. 2020; Bourdin et al. 2022). This could have increased competition with trees, effectively reducing water availability despite irrigation. Finally, emerging research highlights the importance of forest microclimate as a driver of insect activity and diversity (De Frenne et al. 2021; Verdonck et al. 2025; Kerr et al. 2025). Sprinkler irrigation from 2 m high may have altered local microclimate, affected resource distribution, and disrupted host-searching behaviour (Riedener et al. 2013; Li et al. 2024), ultimately influencing herbivory. Our measurements capture only the net outcome of insect feeding on leaves, leaving the underlying antagonistic mechanisms—from leaf to plot scale—unobserved. Future experiments disentangling direct versus indirect, trait-mediated effects of irrigation on herbivores and consumption are needed to clarify these mechanisms and develop a more predictive framework for tree diversity effects on herbivore resistance under contrasting drought regimes.

6.4 Conclusion

Using a unique dataset of repeated measurements spanning 11 years, we evaluated the effects of tree diversity and drought on insect herbivory under field conditions. This long-term perspective enhances the generalisability of our findings by encompassing a wider range of environmental conditions experienced by trees, herbivores, and their natural enemies. However, such an approach limits the ability to isolate specific processes, test mechanistic hypotheses, or directly identify the drivers of observed patterns. Contrary to predictions, we found no consistent overall effect of tree diversity or drought on herbivory. Instead, responses were tree species-specific and context-dependent, varying between attacking herbivores and host trees. This suggests that the net outcome of plant–herbivore interactions along tree diversity gradients reflects the interplay of multiple, potentially antagonistic mechanisms. Our results also highlight the need for caution when generalising from temporally limited experiments. More broadly, they emphasise the value of long-term, replicated studies to capture the variability and contingency inherent to tree–insect interactions under changing environmental conditions.

References

Bansal S (2015) The interactive effects of drought and herbivory on ecophysiology of trees. In: Mahalingam R (ed) Combined stresses in plants: physiological, molecular, and biochemical aspects. Springer International Publishing. https://doi.org/10.1007/978-3-319-07899-1_12

Bognounou F, De Grandpré L, Pureswaran DS, Kneeshaw D (2017) Temporal variation in plant neighborhood effects on the defoliation of primary and secondary hosts by an insect Pest. Ecosphere 8(3):e01759. https://doi.org/10.1002/ecs2.1759

Boppe L (1889) Traité de Sylviculture. Berger-Levrault
Bourdin A, Bord S, Durand J et al (2022) Forest diversity reduces the prevalence of pathogens transmitted by the tick Ixodes Ricinus. Front Ecol Evol 10. https://doi.org/10.3389/fevo.2022.891908
Bourdin A, Augusto L, Joly F-X et al (2025) Context dependency of tree diversity effects on standardized substrates decomposition: role of tree functional composition, Mycorrhizal type and climatic conditions. J Ecol 113(12):3700–3716. https://doi.org/10.1111/1365-2745.70189
Brooks M, Bolker B, Kristensen K et al (2017) glmmTMB: generalized linear mixed models using template model builder. Comprehensive R Archive Network. https://doi.org/10.32614/CRAN.package.glmmTMB
Burnham KP, Anderson DR (2002) Model selection and multimodel inference a practical information-theoretic approach. Springer
Castagneyrol B, Giffard B, Péré C, Jactel H (2013) Plant Apparency, an overlooked driver of associational resistance to insect herbivory. J Ecol (Hoboken) 101(2):418–429. https://doi.org/10.1111/1365-2745.12055
Castagneyrol B, Jactel H, Vacher C, Brockerhoff EG, Koricheva J (2014) Effects of plant phylogenetic diversity on herbivory depend on herbivore specialization. J Appl Ecol (Hoboken) 51(1):134–141. https://doi.org/10.1111/1365-2664.12175
Castagneyrol B, Bonal D, Damien M et al (2017) Bottom-up and top-down effects of tree species diversity on leaf insect herbivory. Ecol Evol (Hoboken) 7(10):3520–3531. https://doi.org/10.1002/ece3.2950
Castagneyrol B, Jactel H, Moreira X (2018a) Anti-herbivore defences and insect herbivory: interactive effects of drought and tree neighbours. J Ecol (Hoboken) 106(5):2043–2057. https://doi.org/10.1111/1365-2745.12956
Castagneyrol B, Moreira X, Jactel H (2018b) Drought and plant neighbourhood interactively determine herbivore consumption and performance. Sci Rep (London) 8(April):5930. https://doi.org/10.1038/s41598-018-24299-x
Castagneyrol B, Giffard B, Valdés-Correcher E, Hampe A (2019) Tree diversity effects on leaf insect damage on Pedunculate oak: the role of landscape context and Forest stratum. Forest Ecol Manag (Amsterdam) 433(February):287–294. https://doi.org/10.1016/j.foreco.2018.11.014
Castagneyrol B, Kozlov M, Poeydebat C, Toigo M, Jactel H (2020) Associational resistance to a pest insect fades with time. J Pest Sci (Heidelberg) 93(1):427–437. https://doi.org/10.1007/s10340-019-01148-y
Corcket E, Alard D, van Halder I et al (2020) Canopy composition and drought shape understorey plant assemblages in a young tree diversity experiment. J Veg Sci 31(5):803–816. https://doi.org/10.1111/jvs.12903
Csóka G, Hirka A, Mutun S et al (2020) Spread and potential host range of the invasive oak lace bug [Corythucha Arcuata (say, 1832) – Heteroptera: Tingidae] in Eurasia. Agric For Entomol 22(1):61–74. https://doi.org/10.1111/afe.12362
Damestoy T, Moreira X, Jactel H, Valdés-Correcher E, Plomion C, Castagneyrol B (2021) Growth and mortality of the oak Processionary moth, Thaumetopoea Processionea, on two oak species: direct and trait-mediated effects of host and neighbour species. Entomol Generalis:13–25. https://doi.org/10.1127/entomologia/2020/1005
de Groot M, Ogris N, Diaci J, Castagneyrol B (2023) When tree diversity does not work: the interacting effects of tree diversity, altitude and amount of spruce on European spruce bark beetle outbreaks. For Ecol Manag 537(June):120952. https://doi.org/10.1016/j.foreco.2023.120952
Fahey C, Burghardt KT, Griffin EA et al (2025) Chemically mediated plant-enemy interactions promote positive biodiversity effects on young tree growth. J Ecol (Hoboken) 113(7):1606–1621. https://doi.org/10.1111/1365-2745.70062
Fernandez-Conradi P, Borowiec N, Capdevielle X et al (2018) Plant neighbour identity and invasive pathogen infection affect associational resistance to an invasive gall wasp. Biol Invasions (Dordrecht) 20(6):1459–1473. https://doi.org/10.1007/s10530-017-1637-4
Field E, Castagneyrol B, Gibbs M et al (2020) Associational resistance to both insect and pathogen damage in mixed forests is modulated by tree neighbour identity and drought. J Ecol 108(4):1511–1522. https://doi.org/10.1111/1365-2745.13397

Frenne D, Pieter JL, Luoto M et al (2021) Forest microclimates and climate change: importance, drivers and future research agenda. Glob Chang Biol 27(11):2279–2297. https://doi.org/10.1111/gcb.15569

Galmán A, Vázquez-González C, Röder G, Castagneyrol B (2022) Interactive effects of tree species composition and water availability on growth and direct and indirect defences in Quercus Ilex. Oikos (5):e09125. https://doi.org/10.1111/oik.09125

Gely C, Laurance SGW, Stork NE (2020) How do herbivorous insects respond to drought stress in trees? Biol Rev 95(2):434–448. https://doi.org/10.1111/brv.12571

Giffard B, Jactel H, Corcket E, Barbaro L (2012) Influence of surrounding vegetation on insect herbivory: a matter of spatial scale and herbivore specialisation. Basic Appl Ecol 13(5):458–465. https://doi.org/10.1016/j.baae.2012.08.004

Grossiord C (2020) Having the right neighbors: how tree species diversity modulates drought impacts on forests. New Phytol 228(1):42–49. https://doi.org/10.1111/nph.15667

Gutbrodt B, Mody K, Dorn S (2011) Drought changes plant chemistry and causes contrasting responses in lepidopteran herbivores. Oikos 120(11):1732–1740. https://doi.org/10.1111/j.1600-0706.2011.19558.x

Guyot V, Castagneyrol B, Vialatte A, Deconchat M, Jactel H (2016) Tree diversity reduces pest damage in mature forests across Europe. Biol Lett (London) 12(4):20151037. https://doi.org/10.1098/rsbl.2015.1037

Guyot V, Jactel H, Imbaud B et al (2019) Tree diversity drives associational resistance to herbivory at both forest edge and interior. Ecol Evol (Hoboken) 9(16):9040–9051. https://doi.org/10.1002/ece3.5450

Haase J, Castagneyrol B, Cornelissen JHC et al (2015) Contrasting effects of tree diversity on young tree growth and resistance to insect herbivores across three biodiversity experiments. Oikos (Hoboken) 124(12):1674–1685. https://doi.org/10.1111/oik.02090

Hakimara M, Despland E (2025) Vertical stratification of leaf physical traits exerts bottom–up pressures on insect herbivory in a sugar maple temperate Forest. Insect Conserv Divers 18(1):1–15. https://doi.org/10.1111/icad.12777

Hartig F (2016) DHARMa: residual diagnostics for hierarchical (multi-level/mixed) regression models. Comprehensive R Archive Network. https://doi.org/10.32614/CRAN.package.DHARMa

Hoch G, Stemmelen A, Eötvös CB et al (2024) Infestation intensity by the invasive oak lace bug, Corythucha Arcuata (say) in mixed and pure oak stands. J Appl Entomol 148(1):26–33. https://doi.org/10.1111/jen.13205

Holopainen JK, Virjamo V, Ghimire RP, Blande JD, Julkunen-Tiitto R, Kivimaenpaa M (2018) Climate change effects on secondary compounds of forest trees in the northern hemisphere. Front Plant Sci 9(October):1445. https://doi.org/10.3389/fpls.2018.01445

Jactel H, Petit J, Desprez-Loustau M-L et al (2012) Drought effects on damage by forest insects and pathogens: a meta-analysis. Glob Chang Biol 18(1):267–276. https://doi.org/10.1111/j.1365-2486.2011.02512.x

Jactel H, Koricheva J, Castagneyrol B (2019) Responses of forest insect pests to climate change: not so simple. Curr Opin Insect Sci (Amsterdam) 35(October):103–108. https://doi.org/10.1016/j.cois.2019.07.010

Jactel H, Moreira X, Castagneyrol B (2021) Tree diversity and forest resistance to insect pests: patterns, mechanisms and prospects. Annu Rev Entomol 66(1):277–296. https://doi.org/10.1146/annurev-ento-041720-075234

Jactel H, Stemmelen A, Branco M et al (2025) More diverse temperate forests are more resistant against non-native insect herbivores. Preprints. https://doi.org/10.22541/au.175318780.05410473/v1

Jia S, Yang X, Castagneyrol B et al (2024) Neighbouring tree effects on leaf herbivory: insect specialisation matters more than host plant leaf traits. J Ecol 112(1):189–199. https://doi.org/10.1111/1365-2745.14228

Jing X, Muys B, Bruelheide H et al (2021) Above- and below-ground complementarity rather than selection drive tree diversity–productivity relationships in European forests. Funct Ecol 35(8):1756–1767. https://doi.org/10.1111/1365-2435.13825

Kambach S, Kuehn I, Castagneyrol B, Bruelheide H (2016) The impact of tree diversity on different aspects of insect herbivory along a global temperature gradient – a meta-analysis. PLoS One (San Francisco) 11(11):e0165815. https://doi.org/10.1371/journal.pone.0165815

Karban R (2011) The ecology and evolution of induced resistance against herbivores. Funct Ecol 25(2):339–347. https://doi.org/10.1111/j.1365-2435.2010.01789.x

Kerr JT, Susan CC et al (2025) Effects of microclimate variation on insect persistence under global change. Nat Rev Biodivers 1(8):532–542. https://doi.org/10.1038/s44358-025-00067-4

Klapwijk MJ, Bjorkman C (2018) Mixed forests to mitigate risk of insect outbreaks. Scand J Forest Res (Oslo) 33(8):772–780. https://doi.org/10.1080/02827581.2018.1502805

Kozlov MV, Zvereva EL (2017) Background insect herbivory: impacts, patterns and methodology. In: Cánovas FM, Lüttge U, Matyssek R (eds) Progress in botany, vol 79. Springer International Publishing. https://doi.org/10.1007/124_2017_4

Leonard SJ, Dirzo R, Eisenhauer N, Rebollo R, Schädler M, Ferlian O (2023) Explaining variation in plant-herbivore associational effects in a tree biodiversity experiment. J Ecol 111(12):2694–2709. https://doi.org/10.1111/1365-2745.14212

Li Q, Lerner G, Bar E, Lewinsohn E, Tas E (2024) Impact of meteorological conditions on the biogenic volatile organic compound (BVOC) emission rate from eastern Mediterranean vegetation under drought. Biogeosciences 21(18):4133–4147. https://doi.org/10.5194/bg-21-4133-2024

Lin P-A, Kansman J, Chuang W-P, Robert C, Erb M, Felton GW (2023) Water availability and plant–herbivore interactions. J Exp Bot 74(9):2811–2828. https://doi.org/10.1093/jxb/erac481

Muiruri EW, Koricheva J (2016) Going undercover: increasing canopy cover around a host tree drives associational resistance to an insect Pest. Oikos 126(3):339–349. https://doi.org/10.1111/oik.03307

Muiruri EW, Barantal S, Iason GR, Salminen J-P, Perez-Fernandez E, Koricheva J (2019) Forest diversity effects on insect herbivores: do leaf traits matter? New Phytol 21(4):2250–2260. https://doi.org/10.1111/nph.15558

Mundim FM, Pringle EG (2018) Whole-plant metabolic allocation under water stress. Front Plant Sci 9(June). https://doi.org/10.3389/fpls.2018.00852

Novotny V, Miller SE, Baje L et al (2010) Guild-specific patterns of species richness and host specialization in plant-herbivore food webs from a tropical Forest. J Anim Ecol 79(6):1193–1203. https://doi.org/10.1111/j.1365-2656.2010.01728.x

Paulin M, Hirka A, Eötvös C, Csaba G, Mikó Á, Csóka G (2020) Known and predicted impacts of the invasive oak lace bug (Corythucha Arcuata) in European oak ecosystems – a review. Folia Oecologica 47(November):131–139. https://doi.org/10.2478/foecol-2020-0015

Plath M, Dorn S, Riedel J, Barrios H, Mody K (2011) Associational resistance and associational susceptibility: specialist herbivores show contrasting responses to tree stand diversification. Oecologia 169(2):477–487. https://doi.org/10.1007/s00442-011-2215-6

Poeydebat C, Jactel H, Moreira X et al (2021) Climate affects neighbour-induced changes in leaf chemical defences and tree diversity–herbivory relationships. Funct Ecol 35(1):67–81. https://doi.org/10.1111/1365-2435.13700

Poorter H, Niinemets Ü, Ntagkas N et al (2019) A meta-analysis of plant responses to light intensity for 70 traits ranging from molecules to whole plant performance. New Phytol 223(3):1073–1105. https://doi.org/10.1111/nph.15754

Rasheed MU, Brosset A, Blande JD (2023) Tree communication: the effects of 'wired' and 'wireless' channels on interactions with herbivores. Curr Forestry Rep 9(1):33–47. https://doi.org/10.1007/s40725-022-00177-8

Riedener E, Rusterholz H-P, Baur B (2013) Effects of different irrigation systems on the biodiversity of species-rich Hay meadows. Agric Ecosyst Environ 164(January):62–69. https://doi.org/10.1016/j.agee.2012.09.020

Rosado-Sánchez S, Parra-Tabla V, Betancur-Ancona D, Moreira X, Abdala-Roberts L (2017) Tree species diversity alters plant defense investment in an experimental forest plantation in southern Mexico. Biotropica 50(2):246–253. https://doi.org/10.1111/btp.12527

Rosado-Sánchez S, Parra-Tabla V, Betancur-Ancona D, Moreira X, Abdala-Roberts L (2018) Effects of tree species diversity on insect herbivory and leaf Defences in *Cordia Dodecandra*: diversity, herbivory, and plant defences. Ecol Entomol. https://doi.org/10.1111/een.12648

Sahu M, Giri AP (2025) Deciphering the role of metabolites and phytohormones in plant resilience to drought and herbivory. Plant Stress (Amsterdam) 15(March):100737. https://doi.org/10.1016/j.stress.2025.100737

Schnabel F, Beugnon R, Yang B, Richter R, Eisenhauer N, Huang Y, Liu X, Wirth C, Cesarz S, Fichtner A, Perles-Garcia MD, Hähn GJA, Härdtle W, Kunz M, Castro Izaguirre NC, Niklaus PA, von Oheimb G, Schmid B, Trogisch S, Wendisch M, Ma K, Bruelheide H (2025) Tree diversity increases forest temperature buffering via enhancing canopy density and structural diversity. Ecol Lett 28(3):e70096. https://doi.org/10.1111/ele.70096

Schuldt A, Scherer-Lorenzen M (2014) Non-native tree species (Pseudotsuga Menziesii) strongly decreases predator biomass and abundance in mixed-species plantations of a tree diversity experiment. For Ecol Manag 327(September):10–17. https://doi.org/10.1016/j.foreco.2014.04.036

Schuldt A, Baruffol M, Böhnke M et al (2010) Tree diversity promotes insect herbivory in subtropical forests of south-East China. J Ecol 98(4):917–926. https://doi.org/10.1111/j.1365-2745.2010.01659.x

Setiawan NN, Vanhellemont M, Baeten L, Dillen M, Verheyen K (2014) The effects of local neighbourhood diversity on Pest and disease damage of trees in a young experimental Forest. For Ecol Manag 334(December):1–9. https://doi.org/10.1016/j.foreco.2014.08.032

Staab M, Schuldt A (2020) The influence of tree diversity on natural enemies – a review of the 'enemies' hypothesis in forests. Curr Fores Rep 6(4):243–259. https://doi.org/10.1007/s40725-020-00123-6

Stemmelen A, Jactel H, Brockerhoff E, Castagneyrol B (2022) Meta-analysis of tree diversity effects on the abundance, diversity and activity of herbivores' enemies. Basic Appl Ecol 58(February):130–138. https://doi.org/10.1016/j.baae.2021.12.003

Stemmelen A, Bastien Castagneyrol, Quentin Ponette et al (2023) Tree diversity reduces co-infestation of douglas fir by two exotic pests and pathogens. NB 84(May):397–413. https://doi.org/10.3897/neobiota.84.94109

Team, R Core (2024) R: a language and environment for statistical computing. R foundation for statistical computing

van Schrojenstein Lantman IM, Hertzog LR et al (2018) Leaf herbivory is more impacted by forest composition than by tree diversity or edge effects. Basic Appl Ecol 29(June):79–88. https://doi.org/10.1016/j.baae.2018.03.006

Vázquez-González C, Pombo-Salinas L, Martín-Cacheda L et al (2022) Effect of water availability on volatile-mediated communication between potato plants in response to insect herbivory. Funct Ecol 36(11):2763–2773. https://doi.org/10.1111/1365-2435.14159

Vazquez-Gonzalez C, Castagneyrol B, Muiruri EW et al (2024) Tree diversity enhances predation by birds but not by arthropods across climate gradients. Ecol Lett 27(5). https://doi.org/10.1111/ele.14427

Vehviläinen H, Koricheva J, Ruohomäki K (2007) Tree species diversity influences herbivore abundance and damage: meta-analysis of long-term forest experiments. Oecologia 152(2):287–298. https://doi.org/10.1007/s00442-007-0673-7

Verdonck S, De Win Y, Nutter T et al (2025) Managing canopy cover to preserve forest microclimate and diverse macroarthropod communities in times of drought. J Appl Ecol 62(10):2582–2593. https://doi.org/10.1111/1365-2664.70137

Wang Z, Feng L, Frew A, Anqi L, Zaipeng Y, Huang Z (2024) Neighbourhood diversity effects on insect herbivory: plant leaf traits mediate associational resistance. J Ecol 112(11):2613–2623. https://doi.org/10.1111/1365-2745.14405

Wein A, Bauhus J, Bilodeau-Gauthier S, Scherer-Lorenzen M, Nock C, Staab M (2016) Tree species richness promotes invertebrate herbivory on congeneric native and exotic tree saplings in a young diversity experiment. PLoS One 11(12):e0168751. https://doi.org/10.1371/journal.pone.0168751

Xiao X, Jorge LR, Volf M et al (2024) The effect of drought-induced leaf traits on Ficus leaf palatability is species specific. Ecosphere (Hoboken) 15(5):e4831. https://doi.org/10.1002/ecs2.4831
Yuan B, Wang B (2025) Insect seedling herbivory is influenced by multiple factors, but the plant apparency theory is more supported than other hypotheses – a case study in a subtropical forest. Forest Ecosyst 12(April):100252. https://doi.org/10.1016/j.fecs.2024.100252
Zeng W, Xiang W, Zhou B et al (2021) Positive tree diversity effect on fine root biomass: via density dependence rather than spatial root partitioning. Oikos 130(1):1–14. https://doi.org/10.1111/oik.07777

Tree Diversity and the Resistance of Urban Forest to Pest Insects: Processes, Patterns, and Prospects

7

Bastien Castagneyrol, Laura Schillé, and Steven Frank

Abstract

Urban trees provide a wide range of vital ecosystem services to city dwellers, yet the delivery of these benefits is increasingly threatened by insect pests. In this chapter, we explore whether the ecological mechanisms known to link tree diversity with forest resistance to insect herbivores in natural and planted forests can be applied to urban environments. We first review recent advances in understanding patterns of insect pest damage on urban trees, along with current baseline knowledge of the abundance, diversity, and spatial distribution of trees in urban forests worldwide. Research has shown that increased tree diversity can interfere with host-searching behavior in forests, thereby reducing the colonization success of insect herbivores. Tree diversity may also influence the expression of anti-herbivore traits in trees and enhance the abundance and activity of natural enemies of herbivores. Here, we review the evidence supporting the role of these mechanisms in shaping tree–herbivore interactions in urban forests. Findings suggest that the effects of tree diversity on herbivores operate mainly through a stronger and more reliable control by natural enemies. In contrast, the effects of tree diversity mediated by altered tree accessibility—whether through resource concentration, apparency, or trait-based mechanisms—remain more controver-

B. Castagneyrol (✉)
BIOGECO, INRAE, University of Bordeaux, Cestas, France
e-mail: bastien.castagneyrol@inra.fr

L. Schillé
Département des sciences biologiques, Université du Québec à Montréal, Montréal, QC, Canada

S. Frank
Department of Entomology and Plant Pathology, North Carolina State University, Raleigh, NC, USA

X. Moreira et al. (eds.), *Effects of Plant Diversity on Plant-Herbivore Interactions*, Ecological Studies 255, https://doi.org/10.1007/978-3-032-26461-9_7

sial. Finally, we identify key conceptual and empirical gaps that hinder the development of a predictive, mechanism-based framework for the effects of tree diversity on insect pests and their damage in cities.

7.1 Introduction

The concept of urban forest encompasses all the trees in a city, whether isolated, planted in rows along streets, or growing in parks or remnant woodlands, to which we can add peripheral forests that are under the influence of the adjacent city (Randrup et al. 2005). It may sound like an oxymoron. The two words, city on the one hand and forest on the other, seem to refer to radically opposing ideas. The city concentrates people and human activities, often at the expense of plants and biodiversity as a whole. In contrast, forests are ecosystems more or less intensively managed that are structured by trees. Urban forests differ markedly from natural and planted forests that have been examined in other chapters of this volume. Yet trees grow in cities and host insect herbivores, some of which are particularly damaging agents capable of altering both tree functioning and that of the broader urban socio-ecosystem. Despite decades of research in natural and planted forests, our understanding of tree–insect interactions in urban forests remains comparatively limited. However, this knowledge gap is increasingly being addressed by a recent surge in ecological studies examining tree–insect interactions within and across cities, both at the level of individual trees or urban forest stands at local or landscape levels. These studies contribute to informing the sustainable management of urban forests.

There are three reasons for dedicating a chapter to urban forest in a book about forest resistance to pests. First, urban forests provide a wide range of essential ecosystem services to city dwellers. Among others, trees cool air temperatures, thereby reducing the urban heat island effect; they buffer air and noise pollution; they reduce rainwater runoff by facilitating water infiltration into the soil; and they provide habitat for tree-associated biodiversity at higher trophic levels (Roy et al. 2012; Mitchell and Devisscher 2022; Savo et al. 2025). The provision of these ecosystem services largely depends, for example, on the spatial distribution of trees in the city (Cimburova and Berghauser Pont 2021; Mitchell and Devisscher 2022; Wang et al. 2024) and on the health of the trees (Suchocka et al. 2023). Second, and related to the previous point, cities often present stressful environments for trees, increasing their vulnerability to both biotic and abiotic stresses, which are often closely intertwined (Frank and Backe 2022; Tabassum et al. 2024). Therefore, surveillance of tree health in cities is key for the sustainability of urban forests and associated services. Last, cities summarize several aspects of global changes, including global warming, habitat and biodiversity loss, and introduction of invasive exotic species (Grimm et al. 2008; Lahr et al. 2018; Theodorou 2022). Within these contexts, the study of tree-herbivore interactions in cities is essential to understand the delivery of ecosystem services to city dwellers as well as provide valuable insights into the future of tree-associated interactions in natural and planted forests outside of cities.

In this chapter, we first provide an overview of herbivory on trees in urban forests. We then detail the typical composition and structure of urban forests. Building upon these two aspects, we discuss whether and how current knowledge on the effect of tree diversity on insect herbivores in "rural" forests can be transposed to urban forests. We specifically examine how tree diversity alters the availability and accessibility of suitable resources for insect herbivores and influences the strength of top-down control exerted by their natural enemies. We conclude by highlighting current gaps in knowledge and identifying priorities for further research.

7.2 Insect Herbivory on Urban Trees

Insect herbivory is determined by a complex interplay between bottom-up forces acting on the insect through its host plant, top-down forces involving its natural enemies, and the abiotic environment determining the activity of the herbivore and the effect of both types of forces. These three components are all influenced by the urban environment (Moreira and Abdala-Roberts 2023) in ways that often make urban trees more vulnerable to insect pests compared to those in natural systems (Sivarajah and Castagneyrol 2024).

Higher temperatures in cities, for example, reduce winter mortality and accelerate the development of herbivorous insects, to the point that the number of generations per year can be higher in urban areas compared to nearby rural environments. Frank and Just (2020) refer to these insects as sleeper species, organisms that might have been innocuous but develop pest-like characteristics when environmental conditions become favorable. This is, for instance, the case of scale insects whose abundance was found to be positively correlated with temperature in the city of Raleigh, North Carolina, and cities throughout the south-eastern United States (Dale and Frank 2014a; Just et al. 2019). She similar response of another scale insect results, in part, from local adaptations to warmer conditions, such as larger size, greater fecundity, and phenological changes (Dale and Frank 2014b; Meineke et al. 2023).

Urban trees also experience chronic water stress as a result of impervious surfaces preventing water from infiltrating into the ground. In turn, water stress triggers changes in plant physiology, generally leading to the accumulation of secondary metabolites and free amino acids in plant cells (Mundim and Pringle 2018; Lin et al. 2023). These changes have direct consequences on insect herbivores. For example, defoliators perform worse on water-stressed plants, whereas moderate drought is often favorable to sap-sucking insects and leaf miners (Jactel et al. 2012; Gely et al. 2020). In addition, several studies reported positive correlation between the percentage of impervious surfaces around trees and the abundance of aphids or sap-sucking herbivores (Koranyi et al. 2020), and negative correlations between impervious surfaces and defoliation by leaf-chewing herbivores (Valdés-Correcher et al. 2022).

Human activities further influence tree-herbivore interactions in cities through diverse sources of pollution that directly affect trees and their insect pests (Schmitt and Burghardt 2021; Theodorou 2022; Moreira and Abdala-Roberts 2023). In the city of Sacramento, California, oak trees growing close to highways with dense vehicle traffic were found to be more defoliated by insect herbivores than control trees in the countryside (Meineke et al. 2023) as a result of changes in leaf traits (Meineke et al. 2023). However, this result does not appear to be generalizable as evidence for chronic pollution on tree-feeding herbivores remains limited (Kozlov 2022) and the response of pest herbivores to air pollution is highly variable (Ryalls et al. 2024). Another source of pollution is Artificial Light At Night (ALAN). ALAN has direct effect on leaf physical and chemical traits, with documented consequences on insect herbivores that differ among species (Cieraad et al. 2023; Cao et al. 2024) and can possibly directly hamper the behavior of insect herbivores (van Langevelde et al. 2017).

Large cities also concentrate fluxes of merchandise and people, thus increasing the probability of accidentally importing pests with plant material, food, or wood products (Branco et al. 2015). Cities also host a large number of exotic trees, thus increasing the probability that an introduced insect finds a suitable host and establishes in cities through a so-called sampling effect (Augustinus et al. 2024). Both mechanisms increase the risk that native trees are attacked by exotic pests, particularly when they share recent co-ancestry with exotic trees.

7.3 α-, β-, and γ-Diversity of Trees in Urban Forests

The composition of urban forests and the spatial distribution of trees in cities are largely shaped by humans, including landscape architects and designers, as well as private landowners. They reflect the historical development of the city, the availability of trees in local nurseries, and the average income of the neighborhood (Avolio et al. 2018). They are deeply influenced by local authorities and the will of residents, at least as much as local abiotic factors (Berland and Elliott 2014; Avolio et al. 2015; Beattie et al. 2025). Regardless of the causes, the diversity and spatial configuration of urban tree communities could have profound consequences on the vulnerability of urban forests to pest insects.

Urban tree diversity is large, skewed, and heterogeneous. At the city level, for example, the number of tree species—γ-diversity—is particularly high in the case of temperate cities. For instance, urban tree inventories conducted in 26 cities in Switzerland totaled 1360 species belonging to 292 genera (Augustinus et al. 2024). In contrast, the surrounding forests only host 76 species belonging to 43 genera (Augustinus et al. 2024). Compiling urban tree inventories from 473 urban areas in 73 countries, Ossola et al. (2020) listed 4734 tree species in cities, which represent 8% of the global tree flora. The large species richness of trees at the city level (i.e., γ-diversity) is mostly driven by exotic species planted as ornamental trees or to adapt urban forests to future climatic conditions, or rapidly recolonizing vacant lots (Riley et al. 2018; Augustinus et al. 2024). These patterns have been less investigated in tropical cities, which are often surrounded by highly diverse natural forests.

Large γ-diversity also hides an uneven distribution of species within urban forests (Fig. 7.1). Many tree species planted in urban areas are drawn from a small selection of species-rich genera (*Eucalyptus*, *Quercus*, *Ficus*, *Acacia*, and *Pinus* for

Fig. 7.1 The diversity of urban contexts in which insect herbivores and their natural enemies interact with trees. (**a**) Isolated trees embedded within highly built environments in Zurich, Switzerland. (Photo credit: Bastien Castagneyrol). (**b**) Monospecific alignments of trees planted along an urban boulevard in Bordeaux, France. (Photo credit: Bastien Castagneyrol). (**c**) Plurispecific groups of trees growing within a park setting in Montreal, Canada. (Photo credit: Laura Schillé)

the five most frequent genera; Ossola et al. 2020). In the city of Montreal, for instance, the three most frequent tree species represent one third of trees found in urban forests; in Paris, the London plane tree *Platanus* × *acerifolia* represents 17% of trees found in urban forests (Muller 2024). These numbers are representative of global patterns (Kendal et al. 2014). Zooming in 63 cities in the USA, McCoy et al.

(2022) found that trees significantly clustered by species, with frequent monospecific patches at the street or neighborhood level. Therefore, the heterogeneous and highly skewed distribution of tree diversity within cities often results in low α-diversity but high β-diversity.

7.4 Tree Diversity and Insect Herbivory in Urban Forests

Most ecological research on the resistance of urban forests to insect herbivores has addressed the specificity of urban trees as compared to trees in natural forests, asking three questions: (i) are urban trees more vulnerable to insect herbivores than forest trees (Schueller et al. 2019; Dobrosavljević et al. 2020); (ii) if so, why (Kozlov et al. 2017; Meineke et al. 2023; Roquero et al. 2024), and (iii) are there differences in the vulnerability of native vs. exotic species (Frank et al. 2019; Laux et al. 2022). Another line of research has investigated the effects of the abundance and spatial distribution of trees in trophic interactions between trees, insect herbivores, and their natural enemies (Christie et al. 2010; Garcia et al. 2022; Stemmelen et al. 2025). Still, tests of the effect of tree diversity on insect herbivory in urban trees remain scarce.

A heuristic rule, known as Santamour's rule (Santamours 2004), suggests that an urban forest that is resistant (i.e., suffers less herbivore pressure) and resilient (i.e., remains functional and recovers functions after attacks) to pests (and diseases) should contain less than 10% of trees of the same species, 20% in the same genus, and 30% to the same family (Kendal et al. 2014). The reason is intuitively simple: Cities that are overstocked with a given tree taxon risk losing a large proportion of their urban forest if a pest specializes in that taxon is introduced (Raupp et al. 2006). The current crisis caused in North America by the spread of the Emerald Ash Borer *Agrilus planipennis* is particularly illustrative of this risk (Poland and McCullough 2006; Kovacs et al. 2010).

This rule is consistent with patterns observed in natural and planted forests. Over the last decades, an accrual of field observations (Guyot et al. 2016; Castagneyrol et al. 2019) and tree diversity experiments (Muiruri and Koricheva 2016; Grossman et al. 2018), latter summarized in meta-analyses (Castagneyrol et al. 2014; Jactel et al. 2021), have confirmed the view that tree diversity promotes forest resistance to insect herbivores. The overall predominance of this so-called associational resistance pattern must not negative the existence of the opposite phenomenon, associational susceptibility, whereby a tree species is more vulnerable to insect herbivores when surrounded by heterospecific neighbors (White and Whitham 2000; Frank 2014; Wein et al. 2016). Known mechanisms explaining the occurrence or relative strength of association resistance and susceptibility involve host finding processes, tree direct and indirect defenses, and the top-down control of herbivores by their enemies, with these mechanisms and outcomes being contingent on herbivore traits such as diet breadth.

In this section, we give a brief overview of the theoretical framework linking tree diversity with insect herbivory (the processes). We transpose each mechanism from forests to cities, discussing how much each one is supported from the literature (the *patterns*). Finally, we identify gaps in knowledge, muse on the reasons explaining the scarcity of research on this topic, and propose strategies to overcome difficulties and filling these gaps (the *prospects*).

7.4.1 Diet Specialization

7.4.1.1 Mechanisms

Diet specialization is frequent in insects (Forister et al. 2015). A common distinction is made between specialist herbivores feeding on a few closely related host plants—most generally belonging to the same genus—and generalist herbivores able to feed on several host plants belonging to different genera or families. For instance, the spongy moth *Lymantria dispar* or the fall webworm *Hyphantria cunea* are typical generalist herbivores defoliating a broad range of deciduous trees. On the contrary, the spruce budworm *Choristoneura fumiferana* specializes on the *Picea* sp. genus despite occurrences on firs (Mattson 1983; Nealis and Régnière 2004).

The most consistent pattern reported in the literature concerning the effect of tree diversity on insect herbivores is that generalist herbivores are less affected by tree diversity than specialists (Wein et al. 2016; Jactel et al. 2021). As for generalists, the most plausible explanations are twofold. First, generalist herbivores recognize a border range of host cues facilitating their establishment on a larger number of hosts (Silva and Clarke 2020). Second, feeding on different hosts may contribute to dilute anti-herbivore defenses and improve the balance of nutrients they ingest (Hägele and Rowell-Rahier 1999; Singer et al. 2002), thus resulting in better performance.

7.4.1.2 Implications for Urban Forests

Whether the same difference between generalist and specialist herbivores is to be expected in urban environments is unsure. On the one hand, because urbanization generally has a negative effect on biodiversity that is more pronounced for specialist organisms, one might expect an overrepresentation of generalist pests in urban areas (Valdés-Correcher et al. 2022). In addition, urban tree diversity is comprised largely of exotic species which are likely to support more generalist than specialist herbivores. From this perspective, tree diversity might have a limited effect on reducing insect herbivore damage or could even exacerbate it by providing generalist herbivores with a wider range of potential hosts. As for specialists, on the other hand, expectations are less clear. A greater tree diversity may reduce the ability of insect specialists to find and reach their host trees (see below), which would lead to associational resistance. In contrast, greater tree diversity may force these herbivores to remain on an individual tree when its neighbors are non-hosts. Overall, associational resistance may simply be less likely to arise or to notice in urban forests if specialist herbivores are scarce.

7.4.2 Host Plant-Finding Processes

Insect herbivores use a combination of physical and chemical cues to locate and orientate toward host trees—being the food resource for the foraging individual or the food of the future larvae after egg deposition by foraging gravidic females—while avoiding non-host trees (Bruce et al. 2005; Carrasco et al. 2015; Silva and Clarke 2020). Tree diversity has the potential to interfere with the emission, diffusion, and use of these cues (Jactel et al. 2021).

7.4.2.1 Mechanisms

For specialist herbivores relying on a limited range of host trees, greater tree diversity reduces the availability of suitable resources for herbivorous insects. This may have two consequences: either the concentration of herbivores—and therefore damage—in patches where their host plant is more concentrated (the resource concentration hypothesis; Sholes 2008; de Groot et al. 2023), or the concentration of herbivores on the fewer available host trees (the resource dilution hypothesis; Otway et al. 2005; Damien et al. 2016; Castagneyrol et al. 2020). Both patterns have been documented (Bognounou et al. 2017), but the former (resource concentration) is the most common (Castagneyrol et al. 2014).

Tree diversity may also reduce host apparency (i.e., the probability a host plant is found) by interfering with host finding mechanisms, making it harder for an insect herbivore foraging for food or egg-laying site to locate and reach suitable host trees. For instance, green leaf volatiles that are typically emitted by broadleaved tree species (non-hosts) were found to have a deterring effect on bark beetles associated with pine host trees (Kerr et al. 2017). Likewise, mixing birches (non-host) with pines (host) was found to lower pine infestation rate by a specialist pest, as long as pines remained smaller than neighboring birches (Damien et al. 2016; Castagneyrol et al. 2020), which the authors interpreted as the disruption of visual host cues by taller non-host neighbors.

Both the resource concentration and resource dilution hypotheses involve density-dependent mechanisms, whereby insect herbivores are attracted to a resource patch or a specific plant individual. Although apparency theory focuses on plant traits, it cannot be fully separated from density-dependent mechanisms, as it requires the integration of a ratio between host- and non-host cues by foraging insect herbivores, ultimately determining insect load on a plant. In tree species-rich patches, a major challenge remains in determining the relative contribution of each species and individual to the total resource perceived by an insect herbivore. Addressing this challenge—and making the effects of tree diversity predictable—will require further conceptual and methodological advancements, particularly in highly diverse and heterogeneous urban forests, where a tree's apparency may be influenced by both its neighbors and human infrastructure.

7.4.2.2 Implications for Urban Forests

The effect of host concentration on insect herbivores in cities has been assessed by a few studies who reported contrasting findings that depended on the herbivore species and the spatial scale at which host density is measured (Sperry et al. 2001). In support of the resource concentration hypothesis, the red gum lerp psyllid *Glycaspis brimblecombei* was found to be more abundant in monospecific patches of its *Eucalyptus* host than in paired isolated trees in the same city in Portugal (Garcia et al. 2022). In contrast, in the city of Bordeaux (South-West France), the infestation of horse chestnut *Aesculus hippocastanum* by the horse chestnut leaf miner *Cameraria ohridella*—a specialist leaf miner—showed no statistical association with *Aesculus* density (Stemmelen et al. 2025).

Air pollutants can modify the odorscape used by insect herbivores to find host trees while avoiding non-hosts (Blande et al. 2010). For instance, tropospheric ozone (O_3) can increase or decrease the production of biogenic volatile organic compounds (BVOCs) emitted by urban trees (Blande 2021; Masui et al. 2022, 2023), thus modifying the way a tree can be perceived by insect herbivores. O_3 further interacts with BVOCs in the atmosphere and alters the composition of the odour blend, therefore reducing the attractivity range of a tree within its immediate vicinity (Fuentes et al. 2016; Masui et al. 2023). Such mechanisms may also be at play in natural and planted forests. However, their importance is expected to be more variable in urban forests. Because different tree species have distinct emission profiles in urban environments (Khedive et al. 2017; Ahn et al. 2022), tree diversity, spatial distribution, and proximity to sources of atmospheric pollution are likely to influence host-searching behavior and, ultimately, insect herbivory. However, research on this topic is still in its infancy, limiting current predictions and practical applications.

7.4.3 The Particular Case of Exotic Tree Species

7.4.3.1 Mechanisms

Exotic trees may have a negative impact on the diversity of native insect herbivores (Schuldt and Scherer-Lorenzen 2014) but not always (Frank et al. 2019). Exotic tree species are less palatable to and host fewer native herbivores (Burghardt et al. 2010). Paradoxally, in some cases, they may also be associated with similar herbivore communities (Frank et al. 2019), many times as a function of their relatedness with native trees. Exotic trees introduced outside of their native range generally escape from their specialized herbivores and incur less herbivory in the introduced range (Branco et al. 2015). The probability they are colonized by native herbivores increases with the area covered by the exotic tree species in the introduced range, as well as with the presence of closely related native tree species (Branco et al. 2015). These effects can be mediated by local tree diversity and species composition (Martinson et al. 2020; Schuldt et al. 2022) and are further influenced by herbivore traits (Wein et al. 2016; Berthelot et al. 2023). Generalist herbivores being particularly prone to switch to introduced exotic tree species are more likely to benefit from tree diversity as compared to specialist herbivores (Hinman et al. 2019; Berthelot et al. 2023; Stazione et al. 2025).

Insect herbivores are also transported across continents, and the amount of damage they may cause to trees is also influenced by tree diversity (Fernandez-Conradi et al. 2018; Jactel et al. 2025). The (fortunately) limited number of examples prevents broad generalization; however, emblematic cases illustrate differences in tree vulnerability to exotic pests with varying diet breadth. In North America, native trees are often more susceptible to specialized exotic insects, such as emerald ash borer which has decimated ash populations, than generalists, such as Asian longhorned beetle (*Anoplophora glabripennis*), which has killed fewer trees and spreads slowly.

7.4.3.2 Implications for Urban Forests

Exotic species are widely used in cities (Silva and Clarke 2020; Chambers-Ostler et al. 2024; Augustinus et al. 2024). They generally suffer less damage from insect herbivores (Matter et al. 2012; Frank 2014; Clem and Held 2018; Laux et al. 2022), which makes them healthier than native species (Matter et al. 2012). Concomitantly, however, they support a lower biodiversity (Helden et al. 2012; Clem and Held 2018; Laux et al. 2022) as they are mostly used by generalist herbivores.

The contribution of exotic tree species to associational effects is unclear. Few authors specifically addressed the response (Frank 2014; Stemmelen et al. 2022b; Schillé et al. 2025) of exotic species to tree diversity or their effect on herbivory as a whole (Clem and Held 2018; Riley et al. 2022; Wilson et al. 2023). On the one hand, the spatial proximity between native and exotic trees may facilitate the spillover of native herbivores onto the later (i.e., associational susceptibility). On the other hand, the presence of exotic trees in the urban landscape could increase associational resistance by diluting herbivore attacks and saturating the environment with non-host visual and chemical cues. Neither mechanism has received strong support. Studies conducted thus far suggest that tree diversity benefits both native and exotic trees to a similar extent (Frank 2014; Stemmelen et al. 2022b; Schillé et al. 2025), just as the presence of exotic trees does not appear to have any influence on the general protective effect of tree diversity against insect herbivores on native trees (Riley et al. 2022; Schillé et al. 2025). For instance, when comparing insect herbivory and the top-down control exerted by insectivorous birds on insect herbivores in 25 urban forest plots varying in tree species diversity and composition in the city of Montreal (Canada), Schille et al. (2025) found no particular relationship between the proportion of native trees and neither herbivory nor predation (Fig. 7.2); see ahead.

7.4.4 Tree Traits, Direct, and Indirect Defenses

7.4.4.1 Mechanisms

The neighbour that a plant has can modify the expression of its physical and chemical traits (Felix et al. 2023, see Chap. 2) that ultimately control its interactions with insect herbivores (Muiruri et al. 2019; Poeydebat et al. 2021). This includes nutritional and defense traits in leaves (Castagneyrol et al. 2018; Muiruri et al. 2019; Jia

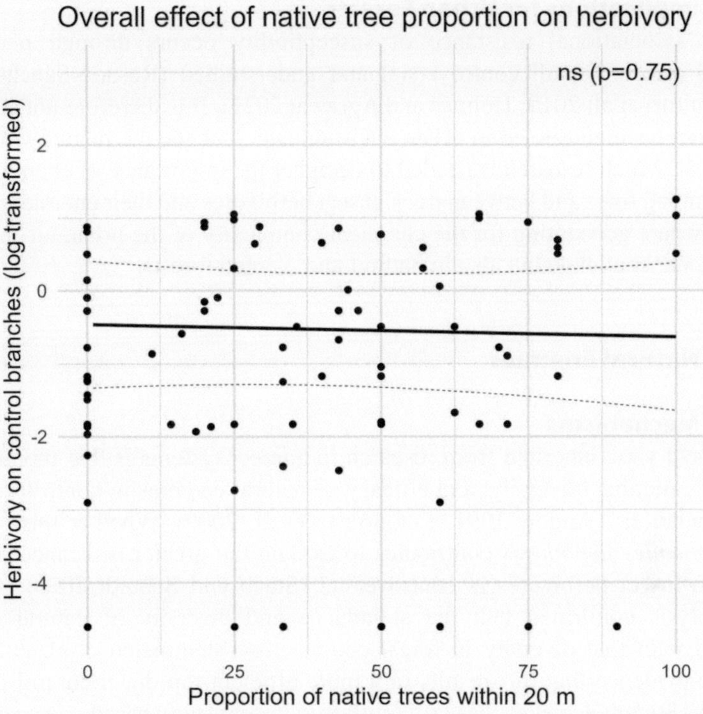

Fig. 7.2 Relationship between the proportion of native trees and the top-down control of insect herbivores by insectivorous birds in the city of Montreal, Canada. Nets were installed on branches of 4 trees focal in 25 urban forest plots in the city of Montreal. The identity of each tree, whether publicly or privately owned, was recorded in a 20 m buffer around each focal tree. The authors compared insect herbivory is the presence (control) or absence (nets) of avian predation at tree level. Each dot represents a single tree. (Data from Schillé et al. 2025)

et al. 2020; Damestoy et al. 2021) as well as the emission of BVOCs (Galmán et al. 2022) that act as kairomones signaling the presence of insect herbivores to natural enemies (Llusià and Peñuelas 2001; Mäntylä et al. 2020; Rasheed et al. 2023). Several studies conducted in tree diversity experiments (Rosado-Sánchez et al. 2017; Castagneyrol et al. 2018; Poeydebat et al. 2021) or natural settings (Castagneyrol et al. 2019; Wang et al. 2022; Jia et al. 2024) reported intraspecific variability in leaf traits that was accounted for by the identity or diversity of a tree's neighbor. However, the relative importance of trait-mediated effects of tree diversity on herbivory is a matter of debate (Jactel et al. 2021). Some authors clearly identified changes in leaf traits as a functional link between tree diversity and herbivory (Muiruri et al. 2019; Wang et al. 2022), whereas others found independent (Rosado-Sánchez et al. 2017; Castagneyrol et al. 2018) or context-dependent (Poeydebat et al. 2021) effects.

7.4.4.2 Implications for Urban Forests

Whether associational resistance of susceptibility occurs through plant trait-mediated effects is still controversial and understudied (Rosado-Sánchez et al. 2017; Muiruri et al. 2019; Holmes and Agrawal 2021). It is therefore not surprising this has not been addressed in urban environments (but see the related work by Li et al. 2025). Much research is needed to decipher the importance of chemical interactions among trees and between trees, insect herbivores and their enemies, not only in cities where accounting for the chemical complexity of the urban air represents an additional challenge but also in natural and planted forests.

7.4.5 Natural Enemies

7.4.5.1 Mechanisms

A long-held view inherited from research in agroecosystems is that monocultures reduce the number, diversity, and efficacy of natural enemies in controlling herbivore populations (Andow 1991; Letourneau et al. 2009). Whether this so-called *natural enemies hypothesis* contributes to explain the greater resistance of mixed forests to insect herbivores is controversial (Staab and Schuldt 2020). A recent meta-analysis confirmed that the abundance and diversity of natural enemies increased with tree diversity in forest ecosystems (Stemmelen et al. 2022a) but found no evidence that this results in a more efficient top-down control of insect herbivory (Stemmelen et al. 2022a). Similarly, a recent study based on experimental tree diversity plots found that bird predation on artificial caterpillars increased with surrounding tree species richness, whereas no such effect was observed for arthropod predators (Vázquez-González et al. 2024).

7.4.5.2 Implications for Urban Forests

Numerous aspects of urbanization might reduce the potential regulation of herbivore damage in trees by predators and parasitoids. First, urban trees are often isolated, and green spaces are intensively managed, leading to reduced structural vegetation complexity (Aronson et al. 2017) and limited habitat availability for natural enemies such as predatory arthropods (Langellotto and Denno 2004; Frey et al. 2018; Melo et al. 2021). In particular, insectivorous birds and bats are negatively affected by habitat fragmentation and isolation but also by a lack of arthropod prey, which is also dependent on plant diversity and structure (Threlfall et al. 2012; Seress and Liker 2015). Second, urbanization generates disturbances that alter both the composition and behavior of predator communities. For instance, predatory arthropods can show strong declines in abundance and diversity due to intensive vegetation management, chemical pollution, and artificial light (Kotze et al. 2022). Parasitoids are particularly sensitive to extreme temperatures associated with urban heat islands, which can impair their survival and disrupt their phenological synchrony with hosts (Hance et al. 2007). Insectivorous birds (Baker et al. 2005; Slabbekoorn and Ripmeester 2008; Aulsebrook et al. 2020) and bats alike (Russo and Ancillotto 2015; Barré et al. 2021) suffer from concurrent disturbances including noise pollution, artificial lighting, and frequent human and domestic animal presence. Finally, the regulation potential of insect herbivores by their natural enemies might be further reduced in cities by the fact that specialist predators are those that are the most adversely impacted by urbanization (Devictor et al. 2007; Callaghan et al. 2019; Korányi et al. 2022).

Tree diversity has the potential to mitigate the negative impacts of urbanization on herbivores' enemies—notably on bird communities (e Silva et al. 2020)—and the regulation of herbivore damage in cities (Stemmelen et al. 2022a). However, available evidence is still scarce and currently inconsistent (Stemmelen et al. 2022a; Wilson et al. 2023; Basile et al. 2025; Schille et al. 2025). A recent study across three Swiss cities found that insectivorous bird abundance increased with canopy cover but decreased with the richness of non-native tree species (Basile et al. 2025). Omnivorous and granivorous birds, which also consume insects during the breeding season, showed more nuanced responses: Omnivores were positively associated with canopy cover and native tree richness, while granivores responded negatively to canopy cover and noise and exhibited mixed responses to different components of tree diversity. In another study conducted in Montreal, Canada, the species richness and functional diversity of insectivorous birds declined along the urbanization gradient. Interestingly, local tree species richness appeared to partially offset this loss by enhancing bird-driven predation on herbivorous insects (Schillé et al. 2025).

7.5 Prospects

Urban forests profoundly differ from rural or natural forests, in several ways which complicate transposing the theoretical background from the latter to urban environments. Further, despite a growing interest in how tree diversity influences insect herbivory in cities, the literature remains sparse and fragmented. Most available studies are correlative, often limited to single-city case studies, and they offer conflicting results. Both abiotic (being climate, light, soil conditions, air quality) and biotic factors (the distribution and diversity of all plants, that of herbivores and their enemies) as well as human disturbance vary widely within and across cities, making broader generalizations about the effects of tree diversity on herbivory in cities still premature.

Current limitations preventing the field from moving forward partly come from the still limited access to relevant environmental data—which can optimistically be solved, partially, by the rapid development of remote sensing and automated classification algorithm technologies—but also from unsuitable theoretical frameworks too coarsely transposed from extra-urban ecosystems, being agroecosystems, and managed or natural forests. On the data side, current understanding of intra-urban heterogeneity in the drivers of insect herbivory still relies on coarse-grained environmental proxies. Existing urban studies have largely relied on easily accessible spatial variables such as tree density and the proportion of impervious surface, which offer only a partial view of how urban environments shape herbivore pressure. Advances in remote sensing technologies and the increasing availability of high-resolution 3D data provide new opportunities to describe urban environments with greater ecological realism. These tools can help better capture the vertical and structural complexity of the urban matrix, which in turn affects microclimates, plant health, and movement patterns of both herbivores and their natural enemies. Unfortunately, even as our ability to detail the aboveground ecosystem increases, our knowledge and ability to monitor, manipulate, and describe soil and belowground processes are likely to remain poor. Soil composition, fertility, moisture, and microbial communities are critical factors influencing tree growth and function and herbivore susceptibility. This remains a fully unexplored area. Additionally, studies usually focus on the canopy layer of trees from public areas. There is an urgent need for comprehensive and exhaustive urban forest inventories accounting for trees in both the public and private domain. While management records often provide detailed accounts of trees in parks and along streets, these overlook trees in private properties, which may constitute a substantial portion of urban forests. Filling this gap, likely with remote sensing and artificial intelligence, will be essential to evaluating how tree diversity is distributed across land ownership types and how it might influence insect herbivore dynamics city-wide.

All this is not to say we do not have a lot of data and knowledge of urban ecosystems and ecological processes that influence herbivory. However, this research has not been thoroughly integrated into existing theoretical models and frameworks. Theoretical research explaining and predicting the effect of urban tree diversity on insect herbivores and herbivory could provide insight and testable hypotheses to improve field research precision and efficiency. Studies usually focus on the canopy layer alone, overlooking the role of understory and ground-level vegetation, which can act as both a reservoir or refuge for herbivores and as critical habitat for predators. For instance, the diversity of nonwoody plants but also their arrangement at small spatial scale were found to be key predictors of the activity of natural enemies (Mitchell et al. 2023a, b). Improving our ability to quantify these fine-scale ecological drivers is essential for advancing mechanistic understanding of tree-herbivore interactions in urban areas.

Likewise, urban ecology has so far relied on a wealth of city-level reports. Many of which serve as instructive case studies. The lack of diversity in case studies implicitly forces us to consider “cities” as coherent and commensurable objects. Yet, urban forms vary greatly across continents and biomes. Without cross-city comparisons that use standardized sampling protocols, it is difficult to determine whether observed associations between tree diversity and herbivory are robust or simply idiosyncratic products of local context. A comparative approach would allow researchers to test the consistency of ecological relationships under similar climatic and ecological conditions but under varying urban designs, historical legacies, and biodiversity baselines. Coordinated cross-continental studies would provide an essential foundation for developing a more coherent theoretical framework linking urban biodiversity to ecosystem functioning via altered species interactions.

References

Ahn J-W, Dinh T-V, Park S-Y et al (2022) Characteristics of biogenic volatile organic compounds emitted from major species of street trees and urban forests. Atmos Pollut Res 13:101470. https://doi.org/10.1016/j.apr.2022.101470

Andow DA (1991) Vegetational diversity and arthropod population response. Annu Rev Entomol 36:561–586

Aronson MF, Lepczyk CA, Evans KL, Goddard MA, Lerman SB, MacIvor JS, Nilon CH, Vargo T (2017) Biodiversity in the city: key challenges for urban green space management. Front Ecol Environ 15:189–196

Augustinus BA, Abegg M, Queloz V, Brockerhoff EG (2024) Higher tree species richness and diversity in urban areas than in forests: implications for host availability for invasive tree pests and pathogens. Landsc Urban Plan 250:105144. https://doi.org/10.1016/j.landurbplan.2024.105144

Aulsebrook AE, Connelly F, Johnsson RD et al (2020) White and Amber light at night disrupt sleep physiology in birds. Curr Biol 30:3657–3663. https://doi.org/10.1016/j.cub.2020.06.085

Avolio M, Pataki DE, Gillespie T, Jenerette GD, McCarthy HR, Pincetl S, Weller-Clarke L (2015) Tree diversity in southern California's urban forest: the interacting roles of social and environmental variables. Front Ecol Evol. in press.

Avolio ML, Pataki DE, Trammell TLE, Endter-Wada J (2018) Biodiverse cities: the nursery industry, homeowners, and neighborhood differences drive urban tree composition. Ecol Monogr 88:259–276

Baker P, Bentley A, Ansell R, Harris S (2005) Impact of predation by domestic cats Felis catus in an urban area. Mammal Rev 35:302–312

Barré K, Spoelstra K, Bas Y et al (2021) Artificial light may change flight patterns of bats near bridges along urban waterways. Anim Conserv 24:259–267. https://doi.org/10.1111/acv.12635

Basile M, Augustinus BA, Brockerhoff EG (2025) Tree cover and geographic origin of tree species drive bird functional groups in urban environments. Wildl Biol:e01446. https://doi.org/10.1002/wlb3.01446

Beattie LH, King G, Hvenegaard G (2025) From private yards to public benefits: how human dimensions shape the urban forest in a small city. Urban For Urban Green 112:128948

Berland A, Elliott GP (2014) Unexpected connections between residential urban forest diversity and vulnerability to two invasive beetles. Landsc Ecol 29:141–152

Berthelot S, Bauhus J, Dormann CF et al (2023) Exotic tree species have consistently lower herbivore load in a cross-Atlantic tree biodiversity experiment. Ecology 104:e4070. https://doi.org/10.1002/ecy.4070

Blande JD (2021) Effects of air pollution on plant–insect interactions mediated by olfactory and visual cues. Curr Opin Environ Sci Health 19:100228. https://doi.org/10.1016/j.coesh.2020.100228

Blande JD, Holopainen JK, Li T (2010) Air pollution impedes plant-to-plant communication by volatiles. Ecol Lett 13:1172–1181. https://doi.org/10.1111/j.1461-0248.2010.01510.x

Bognounou F, De Grandpré L, Pureswaran DS, Kneeshaw D (2017) Temporal variation in plant neighborhood effects on the defoliation of primary and secondary hosts by an insect pest. Ecosphere 8:e01759. https://doi.org/10.1002/ecs2.1759

Branco M, Brockerhoff EG, Castagneyrol B et al (2015) Host range expansion of native insects to exotic trees increases with area of introduction and the presence of congeneric native trees. J Appl Ecol 52:69–77. https://doi.org/10.1111/1365-2664.12362

Bruce TJA, Wadhams LJ, Woodcock CM (2005) Insect host location: a volatile situation. Trends Plant Sci 10:269–274. https://doi.org/10.1016/j.tplants.2005.04.003

Burghardt KT, Tallamy DW, Philips C, Shropshire KJ (2010) Non-native plants reduce abundance, richness, and host specialization in lepidopteran communities. Ecosphere 1:art11. https://doi.org/10.1890/ES10-00032.1

Callaghan CT, Major RE, Wilshire JH et al (2019) Generalists are the most urban-tolerant of birds: a phylogenetically controlled analysis of ecological and life history traits using

a novel continuous measure of bird responses to urbanization. Oikos 128:845–858. https://doi.org/10.1111/oik.06158
Cao Y, Zhang S, Ma K-M (2024) Artificial light at night decreases leaf herbivory in typical urban areas. Front Plant Sci 15. https://doi.org/10.3389/fpls.2024.1392262
Carrasco D, Larsson MC, Anderson P (2015) Insect host plant selection in complex environments. Curr Opin Insect Sci 8:1–7. https://doi.org/10.1016/j.cois.2015.01.014
Castagneyrol B, Jactel H, Vacher C et al (2014) Effects of plant phylogenetic diversity on herbivory depend on herbivore specialization. J Appl Ecol 51:134–141. https://doi.org/10.1111/1365-2664.12175
Castagneyrol B, Jactel H, Moreira X (2018) Anti-herbivore defences and insect herbivory: interactive effects of drought and tree neighbours. J Ecol 106:2043–2057. https://doi.org/10.1111/1365-2745.12956
Castagneyrol B, Giffard B, Valdés-Correcher E, Hampe A (2019) Tree diversity effects on leaf insect damage on pedunculate oak: the role of landscape context and forest stratum. For Ecol Manag 433:287–294. https://doi.org/10.1016/j.foreco.2018.11.014
Castagneyrol B, Kozlov M, Poeydebat C et al (2020) Associational resistance to a pest insect fades with time. J Pest Sci 93:427–437. https://doi.org/10.1007/s10340-019-01148-y
Chambers-Ostler A, Walker H, Doick KJ (2024) The role of the private tree in bringing diversity and resilience to the urban forest. Urban For Urban Green 91:127973. https://doi.org/10.1016/j.ufug.2023.127973
Christie FJ, Cassis G, Hochuli DF (2010) Urbanization affects the trophic structure of arboreal arthropod communities. Urban Ecosyst 13:169–180. https://doi.org/10.1007/s11252-009-0115-x
Cieraad E, Strange E, Flink M et al (2023) Artificial light at night affects plant–herbivore interactions. J Appl Ecol 60:400–410. https://doi.org/10.1111/1365-2664.14336
Cimburova Z, Berghauser Pont M (2021) Location matters. A systematic review of spatial contextual factors mediating ecosystem services of urban trees. Ecosyst Serv 50:101296. https://doi.org/10.1016/j.ecoser.2021.101296
Clem CS, Held DW (2018) Associational interactions between urban trees: are native neighbors better than non-natives? Environ Entomol 47:881–889. https://doi.org/10.1093/ee/nvy071
Dale AG, Frank SD (2014a) The effects of urban warming on herbivore abundance and street tree condition. PLoS One 9:e102996. https://doi.org/10.1371/journal.pone.0102996
Dale AG, Frank SD (2014b) Urban warming trumps natural enemy regulation of herbivorous pests. Ecol Appl 24:1596–1607. https://doi.org/10.1890/13-1961.1
Damestoy T, Moreira X, Jactel H et al (2021) Growth and mortality of the oak processionary moth, Thaumetopoea processionea, on two oak species: direct and trait-mediated effects of host and neighbour species. Entomol Generalis:13–25. https://doi.org/10.1127/entomologia/2020/1005
Damien M, Jactel H, Meredieu C et al (2016) Pest damage in mixed forests: disentangling the effects of neighbor identity, host density and host apparency at different spatial scales. For Ecol Manag 378:103–110. https://doi.org/10.1016/j.foreco.2016.07.025
de Groot M, Ogris N, Diaci J, Castagneyrol B (2023) When tree diversity does not work: the interacting effects of tree diversity, altitude and amount of spruce on European spruce bark beetle outbreaks. For Ecol Manag 537. https://doi.org/10.1016/j.foreco.2023.120952
Devictor V, Julliard R, Couvet D et al (2007) Functional homogenization effect of urbanization on bird communities. Conserv Biol 21:741–751. https://doi.org/10.1111/j.1523-1739.2007.00671.x
Dobrosavljević J, Marković Č, Marjanović M, Milanović S (2020) Pedunculate oak leaf miners' community: urban vs. rural habitat. Forests 11:1300. https://doi.org/10.3390/f11121300
e Silva JLS, de Oliveira MTP, Oliveira W et al (2020) High richness of exotic trees in tropical urban green spaces: reproductive systems, fruiting and associated risks to native species. Urban For Urban Green 50:126659. https://doi.org/10.1016/j.ufug.2020.126659
Felix JA, Stevenson PC, Koricheva J (2023) Plant neighbourhood diversity effects on leaf traits: a meta-analysis. Funct Ecol 37:3150–3163. https://doi.org/10.1111/1365-2435.14441

Fernandez-Conradi P, Borowiec N, Capdevielle X et al (2018) Plant neighbour identity and invasive pathogen infection affect associational resistance to an invasive gall wasp. Biol Invasions 20:1459–1473. https://doi.org/10.1007/s10530-017-1637-4

Forister ML, Novotny V, Panorska AK et al (2015) The global distribution of diet breadth in insect herbivores. PNAS 112:442–447. https://doi.org/10.1073/pnas.1423042112

Frank SD (2014) Bad neighbors: urban habitats increase cankerworm damage to non-host understory plants. Urban Ecosyst 17:1135–1145. https://doi.org/10.1007/s11252-014-0368-x

Frank SD, Backe KM (2022) Effects of urban Heat Islands on temperate forest trees and arthropods. Curr Forestry Rep 9:48–57. https://doi.org/10.1007/s40725-022-00178-7

Frank SD, Just MG (2020) Can cities activate sleeper species and predict future forest pests? A case study of scale insects. Insects 11:142. https://doi.org/10.3390/insects11030142

Frank SD, Backe KM, McDaniel C et al (2019) Exotic urban trees conserve similar natural enemy communities to native congeners but have fewer pests. PeerJ 7:e6531. https://doi.org/10.7717/peerj.6531

Frey D, Vega K, Zellweger F et al (2018) Predation risk shaped by habitat and landscape complexity in urban environments. J Appl Ecol 55:2343–2353. https://doi.org/10.1111/1365-2664.13189

Fuentes JD, Chamecki M, Roulston T et al (2016) Air pollutants degrade floral scents and increase insect foraging times. Atmos Environ 141:361–374. https://doi.org/10.1016/j.atmosenv.2016.07.002

Galmán A, Vázquez-González C, Röder G, Castagneyrol B (2022) Interactive effects of tree species composition and water availability on growth and direct and indirect defences in Quercus ilex. Oikos:e09125. https://doi.org/10.1111/oik.09125

Garcia A, Vaz PG, Franco JC et al (2022) Urban tree isolation affects the abundance of its pests and their natural enemies. Landsc Urban Plan 227:104515. https://doi.org/10.1016/j.landurbplan.2022.104515

Gely C, Laurance SGW, Stork NE (2020) How do herbivorous insects respond to drought stress in trees? Biol Rev 95:434–448. https://doi.org/10.1111/brv.12571

Grimm NB, Faeth SH, Golubiewski NE et al (2008) Global change and the ecology of cities. Science 319:756–760. https://doi.org/10.1126/science.1150195

Grossman JJ, Vanhellemont M, Barsoum N et al (2018) Synthesis and future research directions linking tree diversity to growth, survival, and damage in a global network of tree diversity experiments. Environ Exp Bot 152:68–89. https://doi.org/10.1016/j.envexpbot.2017.12.015

Guyot V, Castagneyrol B, Vialatte A et al (2016) Tree diversity reduces pest damage in mature forests across Europe. Biol Lett 12:20151037. https://doi.org/10.1098/rsbl.2015.1037

Hägele BF, Rowell-Rahier M (1999) Dietary mixing in three generalist herbivores: nutrient complementation or toxin dilution? Oecologia 119:521–533. https://doi.org/10.1007/s004420050815

Hance T, Van Baaren J, Vernon P, Boivin G (2007) Impact of extreme temperatures on parasitoids in a climate change perspective. Annu Rev Entomol 52:107–126. https://doi.org/10.1146/annurev.ento.52.110405.091333

Helden AJ, Stamp GC, Leather SR (2012) Urban biodiversity: comparison of insect assemblages on native and non-native trees. Urban Ecosyst 15:611–624. https://doi.org/10.1007/s11252-012-0231-x

Hinman ED, Fridley JD, Parry D (2019) Plant defense against generalist herbivores in the forest understory: a phylogenetic comparison of native and invasive species. Biol Invasions 21:1269–1281. https://doi.org/10.1007/s10530-018-1898-6

Holmes KD, Agrawal AA (2021) Induced resistance mitigates the effect of plant neighbors on susceptibility to herbivores. Ecosphere 12:e03334. https://doi.org/10.1002/ecs2.3334

Jactel H, Petit J, Desprez-Loustau M-L et al (2012) Drought effects on damage by forest insects and pathogens: a meta-analysis. Glob Chang Biol 18:267–276. https://doi.org/10.1111/j.1365-2486.2011.02512.x

Jactel H, Moreira X, Castagneyrol B (2021) Tree diversity and forest resistance to insect pests: patterns, mechanisms and prospects. Annu Rev Entomol 66:277–296. https://doi.org/10.1146/annurev-ento-041720-075234

Jactel H, Stemmelen A, Branco M et al (2025) More diverse temperate forests are more resistant against non-native insect herbivores

Jia S, Wang X, Yuan Z et al (2020) Tree species traits affect which natural enemies drive the Janzen-Connell effect in a temperate forest. Nat Commun 11:1–9. https://doi.org/10.1038/s41467-019-14140-y

Jia S, Yang X, Castagneyrol B et al (2024) Neighbouring tree effects on leaf herbivory: insect specialisation matters more than host plant leaf traits. J Ecol 112:189–199. https://doi.org/10.1111/1365-2745.14228

Just MG, Dale AG, Long LC, Frank SD (2019) Urbanization drives unique latitudinal patterns of insect herbivory and tree condition. Oikos 128:984–993. https://doi.org/10.1111/oik.05874

Kendal D, Dobbs C, Lohr VI (2014) Global patterns of diversity in the urban forest: is there evidence to support the 10/20/30 rule? Urban For Urban Green 13:411–417. https://doi.org/10.1016/j.ufug.2014.04.004

Kerr JL, Kelly D, Bader MK-F, Brockerhoff EG (2017) Olfactory cues, visual cues, and semiochemical diversity interact during host location by invasive Forest beetles. J Chem Ecol 43:17–25. https://doi.org/10.1007/s10886-016-0792-x

Khedive E, Shirvany A, Assareh MH, Sharkey TD (2017) In situ emission of BVOCs by three urban woody species. Urban For Urban Green 21:153–157. https://doi.org/10.1016/j.ufug.2016.11.018

Koranyi D, Szigeti V, Mezofi L et al (2020) Urbanization alters the abundance and composition of predator communities and leads to aphid outbreaks on urban trees. Urban Ecosyst. https://doi.org/10.1007/s11252-020-01061-8

Korányi D, Egerer M, Rusch A et al (2022) Urbanization hampers biological control of insect pests: a global meta-analysis. Sci Total Environ 834:155396. https://doi.org/10.1016/j.scitotenv.2022.155396

Kotze DJ, Lowe EC, MacIvor JS et al (2022) Urban forest invertebrates: how they shape and respond to the urban environment. Urban Ecosyst 25:1589–1609. https://doi.org/10.1007/s11252-022-01240-9

Kovacs KF, Haight RG, McCullough DG et al (2010) Cost of potential emerald ash borer damage in U.S. communities, 2009–2019. Ecol Econ 69:569–578. https://doi.org/10.1016/j.ecolecon.2009.09.004

Kozlov MV (2022) Population dynamics of herbivorous insects in polluted landscapes. Curr Opin Insect Sci 54:100987. https://doi.org/10.1016/j.cois.2022.100987

Kozlov MV, Lanta V, Zverev V et al (2017) Decreased losses of woody plant foliage to insects in large urban areas are explained by bird predation. Glob Chang Biol 23:4354–4364. https://doi.org/10.1111/gcb.13692

Lahr EC, Dunn RR, Frank SD (2018) Variation in photosynthesis and stomatal conductance among red maple (Acer rubrum) urban planted cultivars and wildtype trees in the southeastern United States. PLoS One 13:e0197866. https://doi.org/10.1371/journal.pone.0197866

Langellotto GA, Denno RF (2004) Responses of invertebrate natural enemies to complex-structured habitats: a meta-analytical synthesis. Oecologia 139:1–10. https://doi.org/10.1007/s00442-004-1497-3

Laux M, Lv H, Entling MH et al (2022) Native pedunculate oaks support more biodiversity than non-native oaks, but non-native oaks are healthier than native oaks: a study on street and park trees of a city br. Sci Total Environ 853:158603. https://doi.org/10.1016/j.scitotenv.2022.158603

Letourneau DK, Jedlicka JA, Bothwell SG, Moreno CR (2009) Effects of natural enemy biodiversity on the suppression of arthropod herbivores in terrestrial ecosystems. Annu Rev Ecol Evol Syst 40:573–592. https://doi.org/10.1146/annurev.ecolsys.110308.120320

Li J, Hu Z, Wei H et al (2025) The influence of neighborhood competition on the morphological traits of urban individual tree landscape in Beijing. Urban For Urban Green 112:129006. https://doi.org/10.1016/j.ufug.2025.129006

Lin P-A, Kansman J, Chuang W-P et al (2023) Water availability and plant–herbivore interactions. J Exp Bot 74:2811–2828. https://doi.org/10.1093/jxb/erac481

Llusià J, Peñuelas J (2001) Emission of volatile organic compounds by apple trees under spider mite attack and attraction of predatory mites. Exp Appl Acarol 25:65–77. https://doi.org/10.1023/A:1010659826193

Mäntylä E, Kipper S, Hilker M (2020) Insectivorous birds can see and smell systemically herbivore-induced pines. Ecol Evol 10:9358–9370. https://doi.org/10.1002/ece3.6622

Martinson HM, Raupp MJ, Frank SD (2020) How urban forest composition shapes the structure and function of arthropod communities. In: Urban ecology: its nature and challenges, pp 15–36

Masui N, Agathokleous E, Tani A et al (2022) Plant-insect communication in urban forests: similarities of plant volatile compositions among tree species (host vs. non-host trees) for alder leaf beetle Agelastica coerulea. Environ Res 204:111996. https://doi.org/10.1016/j.envres.2021.111996

Masui N, Shiojiri K, Agathokleous E et al (2023) Elevated O3 threatens biological communications mediated by plant volatiles: a review focusing on the urban environment. Crit Rev Environ Sci Technol 53:1982–2001. https://doi.org/10.1080/10643389.2023.2202105

Matter SF, Brzyski JR, Harrison CJ et al (2012) Invading from the garden? A comparison of leaf herbivory for exotic and native plants in natural and ornamental settings. Insect Sci 19:677–682. https://doi.org/10.1111/j.1744-7917.2012.01524.x

Mattson WJ (1983) Spruce budworm (Choristoneura fumiferana) performance in relation to foliar chemistry of its host plants. In: Talerico RL, Montgomery M (eds) Proceedings, forest defoliator – host interactions: acomparison between gypsy moth and spruce budworms. US Department of Agriculture, Forest Service, Northeastern Forest Experiment Station, New Haven, pp 55–65

McCoy DE, Goulet-Scott B, Meng W et al (2022) Species clustering, climate effects, and introduced species in 5 million city trees across 63 US cities. elife 11:e77891. https://doi.org/10.7554/eLife.77891

Meineke EK, Eng DS, Karban R (2023) Vehicle pollution is associated with elevated insect damage to street trees. J Appl Ecol 60:263–277. https://doi.org/10.1111/1365-2664.14328

Melo TS, Moreira EF, Lopes MVA et al (2021) Influence of urban landscape on ants and spiders richness and composition in forests. Neotrop Entomol 50:32–45. https://doi.org/10.1007/s13744-020-00824-4

Mitchell MGE, Devisscher T (2022) Strong relationships between urbanization, landscape structure, and ecosystem service multifunctionality in urban forest fragments. Landsc Urban Plan 228:104548. https://doi.org/10.1016/j.landurbplan.2022.104548

Mitchell JC, D'Amico V III, Trammell TLE, Frank SD (2023a) Nonnative plant invasion increases urban vegetation structure and influences arthropod communities. Divers Distrib 29:1263–1277. https://doi.org/10.1111/ddi.13755

Mitchell JC, D'Amico V, Trammell TLE, Frank SD (2023b) Carabid specialists respond differently to nonnative plant invasion in urban forests. Urban Ecosyst 26:377–393. https://doi.org/10.1007/s11252-022-01323-7

Moreira X, Abdala-Roberts L (2023) Linking herbivory and ecosystem services in urban forests. Trends Plant Sci 28:139–141. https://doi.org/10.1016/j.tplants.2022.10.006

Muiruri EW, Koricheva J (2016) Going undercover: increasing canopy cover around a host tree drives associational resistance to an insect pest. Oikos 126:339–349. https://doi.org/10.1111/oik.03307

Muiruri EW, Barantal S, Iason GR et al (2019) Forest diversity effects on insect herbivores: do leaf traits matter? New Phytol 21:2250–2260. https://doi.org/10.1111/nph.15558

Muller S (2024) La diversité des forêts urbaines. In: De l'arbre en ville à la forêt urbaine, 1st edn. Editions Quae, Paris

Mundim FM, Pringle EG (2018) Whole-plant metabolic allocation under water stress. Front Plant Sci 9. https://doi.org/10.3389/fpls.2018.00852

Nealis VG, Régnière J (2004) Insecthost relationships influencing disturbance by the spruce budworm in a boreal mixedwood forest. Can J For Res 34:1870–1882. https://doi.org/10.1139/x04-061

Ossola A, Hoeppner MJ, Burley HM et al (2020) The global urban tree inventory: a database of the diverse tree flora that inhabits the world's cities. Glob Ecol Biogeogr 29:1907–1914. https://doi.org/10.1111/geb.13169

Otway SJ, Hector A, Lawton JH (2005) Resource dilution effects on specialist insect herbivores in a grassland biodiversity experiment. J Anim Ecol 74:234–240. https://doi.org/10.1111/j.1365-2656.2005.00913.x

Poeydebat C, Jactel H, Moreira X et al (2021) Climate affects neighbour-induced changes in leaf chemical defences and tree diversity–herbivory relationships. Funct Ecol 35:67–81. https://doi.org/10.1111/1365-2435.13700

Poland TM, McCullough DG (2006) Emerald ash borer: invasion of the urban forest and the threat to North America's ash resource. J For 104:118–124. https://doi.org/10.1093/jof/104.3.118

Randrup TB, Konijnendijk C, Dobbertin MK, Prüller R (2005) The concept of urban forestry in Europe. In: Urban forests and trees. Springer, Berlin, Heidelberg, pp 9–21

Rasheed MU, Brosset A, Blande JD (2023) Tree communication: the effects of "wired" and "wireless" channels on interactions with herbivores. Curr Forestry Rep 9:33–47. https://doi.org/10.1007/s40725-022-00177-8

Raupp MJ, Cumming AB, Raupp EC (2006) Street tree diversity in eastern North America and its potential for tree loss to exotic borers. Arboricul Urban Forestry 32:297–304. https://doi.org/10.48044/jauf.2006.038

Riley CB, Herms DA, Gardiner MM (2018) Exotic trees contribute to urban forest diversity and ecosystem services in inner-city Cleveland, OH. Urban For Urban Green 29:367–376. https://doi.org/10.1016/j.ufug.2017.01.004

Riley CB, Raupp MJ, Fite KL et al (2022) Woody plant biodiversity explains arthropod pest management interventions in residential landscapes. Urban For Urban Green 67:127439. https://doi.org/10.1016/j.ufug.2021.127439

Roquero JO, Lidasan AK, Balasa NKB et al (2024) Predation risk of caterpillar prey is shaped by arthropods and urbanisation in an urban-agricultural landscape: a common garden experiment. Urban Ecosyst. https://doi.org/10.1007/s11252-024-01587-1

Rosado-Sánchez S, Parra-Tabla V, Betancur-Ancona D et al (2017) Tree species diversity alters plant defense investment in an experimental forest plantation in southern Mexico. Biotropica 50:246–253. https://doi.org/10.1111/btp.12527

Roy S, Byrne J, Pickering C (2012) A systematic quantitative review of urban tree benefits, costs, and assessment methods across cities in different climatic zones. Urban For Urban Green 11:351–363. https://doi.org/10.1016/j.ufug.2012.06.006

Russo D, Ancillotto L (2015) Sensitivity of bats to urbanization: a review. Mamm Biol 80:205–212. https://doi.org/10.1016/j.mambio.2014.10.003

Ryalls JMW, Bishop J, Mofikoya AO et al (2024) Air pollution disproportionately impairs beneficial invertebrates: a meta-analysis. Nat Commun 15:5447. https://doi.org/10.1038/s41467-024-49729-5

Santamours FS (2004) Trees for urban planting: diversity uniformity, and common sense. In: Elevitch C (ed) The overstory book: cultivating connections with trees. Springer, pp 396–399

Savo V, D'Amato L, Bartoli F et al (2025) Evaluation of main regulating, provisioning, and supporting ecosystem services of urban street trees: a literature review. Ecosyst Serv 71:101690. https://doi.org/10.1016/j.ecoser.2024.101690

Schillé L, Paquette A, Marcotte G et al (2025) Urban tree diversity fosters bird insectivory despite a loss in bird diversity with urbanization. Landsc Urban Plan 256:105274. https://doi.org/10.1016/j.landurbplan.2024.105274

Schille L, Paquette A, Marcotte G et al (2025) Urban tree diversity fosters bird insectivory despite a loss in bird diversity with urbanization. Landsc Urban Plan 256. https://doi.org/10.1016/j.landurbplan.2024.105274

Schmitt L, Burghardt KT (2021) Urbanization as a disrupter and facilitator of insect herbivore behaviors and life cycles. Curr Opin Insect Sci 45:97–105. https://doi.org/10.1016/j.cois.2021.02.016

Schueller SK, Paul S, Payer N et al (2019) Urbanization decreases the extent and variety of leaf herbivory for native canopy tree species Quercus rubra, Quercus alba, and Acer saccharum. Urban Ecosyst 22:907–916. https://doi.org/10.1007/s11252-019-00866-6

Schuldt A, Scherer-Lorenzen M (2014) Non-native tree species (Pseudotsuga menziesii) strongly decreases predator biomass and abundance in mixed-species plantations of a tree diversity experiment. For Ecol Manag 327:10–17
Schuldt A, Huke P, Glatthorn J et al (2022) Tree mixtures mediate negative effects of introduced tree species on bird taxonomic and functional diversity. J Appl Ecol 59:3049–3060. https://doi.org/10.1111/1365-2664.14300
Seress G, Liker A (2015) Habitat urbanization and its effects on birds. Acta Zool Acad Sci H 61:373–408. https://doi.org/10.17109/AZH.61.4.373.2015
Sholes ODV (2008) Effects of associational resistance and host density on woodland insect herbivores. J Anim Ecol 77:16–23. https://doi.org/10.1111/j.1365-2656.2007.01317.x
Silva R, Clarke AR (2020) The "sequential cues hypothesis": a conceptual model to explain host location and ranking by polyphagous herbivores. Insect Sci 27:1136–1147. https://doi.org/10.1111/1744-7917.12719
Singer MS, Bernays EA, Carrière Y (2002) The interplay between nutrient balancing and toxin dilution in foraging by a generalist insect herbivore. Anim Behav 64:629–643. https://doi.org/10.1006/anbe.2002.3082
Sivarajah S, Castagneyrol B (2024) La santé des arbres urbains: menaces et prévention? In: De l'arbre en ville à la forêt urbaine, 1st edn. Editions Quae, Paris, p 188
Slabbekoorn H, Ripmeester EAP (2008) Birdsong and anthropogenic noise: implications and applications for conservation. Mol Ecol 17:72–83. https://doi.org/10.1111/j.1365-294X.2007.03487.x
Sperry CE, Chaney WR, Shao G, Sadof CS (2001) Effects of tree density, tree species diversity, and percentage of hardscape on three insect pests of honeylocust. Arboricul Urban Forestry 27:263–271. https://doi.org/10.48044/jauf.2001.028
Staab M, Schuldt A (2020) The influence of tree diversity on natural enemies – a review of the "enemies" hypothesis in forests. Curr Forestry Rep. https://doi.org/10.1007/s40725-020-00123-6
Stazione L, Corley JC, Allison JD et al (2025) Novel associations among insect herbivores and trees: patterns of occurrence and damage on pines and eucalypts. Ecol Appl 35:e70018. https://doi.org/10.1002/eap.70018
Stemmelen A, Jactel H, Brockerhoff E, Castagneyrol B (2022a) Meta-analysis of tree diversity effects on the abundance, diversity and activity of herbivores' enemies. Basic Appl Ecol 58:130–138. https://doi.org/10.1016/j.baae.2021.12.003
Stemmelen A, Paquette A, Benot M-L et al (2022b) Insect herbivory on urban trees: complementary effects of tree neighbours and predation. Peer Commun J 2. https://doi.org/10.24072/pcjournal.106
Stemmelen A, Jactel H, Castagneyrol B (2025) Tree diversity and density affect injuries caused by the invasive pest *Cameraria ohridella* in urban areas. Urban For Urban Green 108:128820. https://doi.org/10.1016/j.ufug.2025.128820
Suchocka M, Heciak J, Błaszczyk M et al (2023) Comparison of ecosystem services and replacement value calculations performed for urban trees. Ecosyst Serv 63:101553. https://doi.org/10.1016/j.ecoser.2023.101553
Tabassum S, Manea A, Leishman MR (2024) Limiting the impact of insect pests on urban trees under climate change. Urban For Urban Green 94:128246. https://doi.org/10.1016/j.ufug.2024.128246
Theodorou P (2022) The effects of urbanisation on ecological interactions. Curr Opin Insect Sci 52:100922. https://doi.org/10.1016/j.cois.2022.100922
Threlfall CG, Law B, Banks PB (2012) Influence of landscape structure and human modifications on insect biomass and bat foraging activity in an urban landscape. PLoS One 7:e38800. https://doi.org/10.1371/journal.pone.0038800
Valdés-Correcher E, Popova A, Galmán A et al (2022) Herbivory on the pedunculate oak along an urbanization gradient in Europe: effects of impervious surface, local tree cover, and insect feeding guild. Ecol Evol 12:e8709. https://doi.org/10.1002/ece3.8709
van Langevelde F, van Grunsven RHA, Veenendaal EM, Fijen TPM (2017) Artificial night lighting inhibits feeding in moths. Biol Lett 13:20160874. https://doi.org/10.1098/rsbl.2016.0874

Vázquez-González C, Castagneyrol B, Muiruri EW, Barbaro L, Abdala-Roberts L, Barsoum N, Fruend J, Glynn C, Jactel H, Mcshea WJ, Mereu S, Mooney KA, Morillas L, Nock CA, Paquette A, Parker JD, Parker WC, Roales J, Scherer-Lorenzen M, Schuldt A, Verheyen K, Weih M, Yang B, Koricheva J (2024) Tree diversity enhances predation by birds but not by arthropods across climate gradients. Ecol Lett 27(5):e14427

Wang B, Tian C, Liang Y (2022) Leaf traits-mediated effects of tree diversity on insect herbivory on *Populus laurifolia* in a riparian forest ecosystem. For Ecol Manag 504:119777. https://doi.org/10.1016/j.foreco.2021.119777

Wang C, Jin J, Davies C, Chen WY (2024) Urban forests as nature-based solutions: a comprehensive overview of the National Forest City Action in China. Curr For Rep. https://doi.org/10.1007/s40725-024-00213-9

Wein A, Bauhus J, Bilodeau-Gauthier S et al (2016) Tree species richness promotes invertebrate herbivory on congeneric native and exotic tree saplings in a young diversity experiment. PLoS One 11:e0168751. https://doi.org/10.1371/journal.pone.0168751

White JA, Whitham TG (2000) Associational susceptibility of cottonwood to a box elder herbivore. Ecology 81:1795–1803

Wilson CJ, Backe KM, Just MG et al (2023) Tree species richness around urban red maples reduces pest density but does not enhance biological control. Urban For Urban Green 88:128093. https://doi.org/10.1016/j.ufug.2023.128093

Plant Diversity Effects on Ants: A Key but Overlooked Player in Multitrophic Interactions

8

Xoaquín Moreira and Luis Abdala-Roberts

Abstract

Plant diversity exerts strong control over trophic interactions and community dynamics. While the effects of plant diversity on insect herbivores and predators have been widely studied, less attention has been given to ants. Ants are widespread and ecologically important omnivorous predators and mutualists that contribute significantly to trophic dynamics. They are highly responsive to both biotic and abiotic habitat features influenced by plant diversity. This chapter reviews current evidence on the effects of plant diversity on ant communities and their interactions with plants, herbivores, and other predators. We begin by synthesizing findings from experimental and observational studies, focusing on the effects of plant diversity on ant community structure. We then examine how plant diversity influences ant-associated interactions—specifically, ant-herbivore and ant-predator antagonisms, as well as ant-plant and ant-herbivore mutualisms that modulate those antagonisms. Following this, we synthesize major patterns of ant-mediated top-down control as a function of plant diversity and compare ant responses and their trophic effects with those of other predator groups. Finally, we identify promising lines of inquiry for future research on how plant diversity affects ant communities. Overall, this chapter emphasizes the need to better acknowledge and more fully integrate ants into research on plant diversity and trophic interactions. Addressing this missing piece can enhance our understanding of how plant diversity shapes multitrophic interactions and, in turn, how these interactions influence ecosystem functioning—particularly in the context

X. Moreira (✉)
Misión Biológica de Galicia (MBG-CSIC), Pontevedra, Spain

L. Abdala-Roberts
Universidad Nacional Autónoma de México, Escuela Nacional de Estudios Superiores-Unidad Mérida, Ucú, Yucatán, Mexico

X. Moreira et al. (eds.), *Effects of Plant Diversity on Plant-Herbivore Interactions*, Ecological Studies 255, https://doi.org/10.1007/978-3-032-26461-9_8

of global change affecting biodiversity and its ecological consequences in both natural and managed systems.

8.1 Introduction

Plant diversity exerts strong controls over ecosystem functioning (Hooper et al. 2005; Cardinale et al. 2011; Letourneau et al. 2011; Tilman et al. 2014), and evidence is mounting that these effects are mediated by changes in plant–animal interactions and community structure (Moreira et al. 2016; Barnes et al. 2020; Li et al. 2023). Within this body of work, plant diversity effects on herbivory (Moreira et al. 2016; Jactel et al. 2021) and predation (Letourneau et al. 2011; Staab and Schuldt 2020; Stemmelen et al. 2022) are well established (see global synthesis by Wan et al. 2020). Studies manipulating plant diversity have shown that it positively influences herbivore and predator diversity (Haddad et al. 2009; Moreira et al. 2016), often triggering trophic cascades where increased predator pressure reduces herbivore abundance, thereby releasing plants from herbivory and boosting primary productivity (Letourneau et al. 2009; Moreira et al. 2016; Stemmelen et al. 2022). However, multitrophic interaction responses to plant diversity are highly variable and depend on species traits and community structure across trophic levels (Duffy et al. 2007; Letourneau et al. 2011; Abdala-Roberts and Moreira 2024). Accordingly, a central aim of plant diversity research has been to understand the drivers behind variability in the direction and magnitude of predator responses to plant diversity (Haddad et al. 2009; Scherber et al. 2010; Moreira et al. 2016; Holmes and Blubaugh 2023) and the consequences for food web structure, trophic control, and, ultimately, ecosystem functioning (Abdala-Roberts and Moreira 2025).

Increased predator and parasitoid abundance in diverse plant communities is attributed to expanded niche space (Letourneau et al. 2009). However, the mechanisms underlying this condition—and their implications for food web and ecosystem dynamics—remain in most cases poorly understood (Letourneau et al. 2009; Moreira et al. 2016; Stemmelen et al. 2022). For instance, we lack detailed knowledge of which interrelated components of predator communities are most responsive to plant diversity and most strongly influence top-down control (Nell et al. 2018). Similarly, the specific traits of plants, prey, or predators that mediate plant diversity effects on multitrophic interactions are often not explicitly studied (e.g. plant chemistry or predator traits such as hunting mode or mobility; reviewed by Snyder 2019; Abdala-Roberts and Moreira 2024). Despite these shortfalls, empirical studies have shed important insights into potential mechanisms that serve as guide. For example, a major driver of plant diversity effects on predator diversity and foraging behaviour is thought to be increased structural complexity (Lawton 1983; Andow 1991; Denno et al. 2005), via increased availability of refugia and dampening of negative interactions among predators (e.g. Nell and Mooney 2019; Abdala-Roberts and Moreira 2025). Plant diversity can also increase the availability or diversity of plant-based food resources—such as extrafloral nectar—with strong positive effects on generalist (omnivorous) predators (Jones et al. 2017;

Abdala-Roberts and Moreira 2025). Likewise, higher plant diversity also leads to increased prey abundance and diversity, indirectly driving more abundant and species-rich predator communities (Moreira et al. 2016; Schuldt et al. 2019). Robustly testing these multitrophic mechanisms and disentangling the relative importance—e.g. across different predator groups—remains a critical challenge for elucidating plant diversity controls over ecosystem functioning via trophic interactions.

Within this body of research, the effects of plant diversity on invertebrate predators have been intensively studied (reviewed by Letourneau et al. 2011; Stemmelen et al. 2022). Examples include numerous groups such as spiders (Schuldt and Staab 2015; Ezquivel-Gómez et al. 2016), predatory beetles (Zou et al. 2013; Bustos-Segura et al. 2017), predatory bugs (Finke and Denno 2002), predatory mites (Daud 2018), and parasitoids (Herz and Heitland 2005; Abdala-Roberts et al. 2016), among others (reviewed by Letourneau et al. 2009; Staab and Schuldt 2020). In contrast, and despite their high abundance and key ecological roles, ants have received comparatively less attention (but see Moreira et al. 2012; Achury et al. 2022). Ants are clearly one of the most ecologically important and globally widespread predator groups (Hölldobler and Wilson 1990; Lach et al. 2010), contributing significantly to herbivore suppression (Heil and Mckey 2003; Stüber et al. 2021). Plant diversity can influence ant community structure and associated interactions through multiple pathways. As for other predator groups, these include increased vegetation complexity, which may facilitate escape from other predators (e.g. Hertzog et al. 2017), and greater availability or diversity of prey or mutualistic partners (e.g. Staab et al. 2021), the latter via plant-based food resources such as extrafloral nectar (e.g. Staab et al. 2017; Staab et al. 2023) (Fig. 8.1a, b) or sap-feeding herbivores (e.g. Moreira et al. 2012; Staab et al. 2015) (Fig. 8.2a, b). In addition, plant diversity also affects habitat features such microclimatic conditions or the availability of nesting sites crucial for colony establishment and success, including both canopy and leaf litter habitats (Achury et al. 2022). Some of these drivers will likely be more important for ants than for other predator groups (e.g. extrafloral nectar) or even unique to ants (ant-sap-feeding herbivore interactions) but for the most part remain poorly

Fig. 8.1 (**a**) An ant (*Ectatomma tuberculatum*) feeding on an extrafloral nectary of *Peixotoa tomentosa* in the Brazilian Cerrado. (**b**) *Ectatomma opaciventre* ants feeding on pericarpial nectaries of *Palicourea rigida* in the Brazilian Cerrado. (Photo credits: Vítor M. Costa-Silva)

Fig. 8.2 (**a**) An Argentine ant (*Linepithema humile*) tending a colony of cotton aphids (*Aphis gossypii*) feeding on *Baccharis salicifolia* in southern California (Photo credit: Kailen A. Mooney). (**b**) Ants tending *Cinara pinimaritimae* aphids on *Pinus pinaster* in a plant species diversity experiment in Spain. (Photo credit: Xoaquín Moreira)

understood (Moreira et al. 2016; Staab and Schuldt 2020). Accordingly, this gap hinders our ability to predict how plant diversity influences ant communities and ant-mediated interactions dictating top-down control (Staab and Schuldt 2020; Abdala-Roberts and Moreira 2025). Addressing this gap represents a key priority in plant diversity research, especially given growing evidence that ant responses to plant diversity are strong and widespread (Moreira et al. 2012; Staab et al. 2021, 2023), and can exert substantial influence on food web and ecosystem dynamics, including crop agroecosystems (Anjos et al. 2022; Sparer et al. 2024) and forestry plantations (Staab and Schuldt 2020; Stemmelen et al. 2022).

In this chapter, we review research on the effects of plant diversity on ants, focusing on their roles as omnivorous predators engaging in diverse interactions with plants, herbivores, and predators. We begin by briefly summarizing empirical evidence on plant diversity effects on ant communities, namely, abundance, diversity, and community composition. We then examine studies that have investigated the effects of plant diversity on ant–herbivore (both antagonistic and mutualistic) and ant–predator interactions, including work on plant-derived foods such as extrafloral nectar and floral resources. Following from this, we synthesize main patterns of ant-mediated top-down control as a function of plant diversity and then discuss how ant responses and their trophic effects compare to those of other predator groups, highlighting unique features (traits, interactions) that set them aside relative to other predator groups and need to be considered in testing their responses to plant diversity. Finally, we identify key research gaps in our understanding of plant diversity effects on ants, which we believe could help guide research. In addressing these points, the overall goal of this chapter is to underscore the importance of ants in plant diversity–multitrophic interaction research and to call for increased recognition about their role in shaping plant diversity effects on food web and ecosystem functioning.

8.2 Review of Plant Diversity Effects on Ants

In this section, we synthesize theoretical and empirical research examining plant diversity effects ant communities and ant-associated interactions acting through various bottom-up mechanisms linked to plant traits, particularly plant foods. We cover experimental studies that manipulate plant diversity to evaluate ant responses, as well as observational studies conducted along natural gradients of plant diversity. Other plant traits or community features—such as structural complexity—are likely important but have been virtually unstudied and are therefore not covered in this section (but see *Future directions*).

8.2.1 Overview of Plant Diversity Effects on Ant Communities

Research from the BEF-China tree diversity experiment—conducted across a gradient from monocultures to 24-species mixtures in subtropical forests—provides some of the best evidence that plant diversity boosts canopy-associated ant species richness and abundance (Staab et al. 2014, 2017; Schuldt et al. 2017). Likewise, follow-up work in this system by Fornoff et al. (2019) also reported that tree diversity promotes the abundance and diversity of predatory ants. In addition, studies in the same system have found that tree diversity positively influences with leaf litter ant abundance and species richness and phylogenetic diversity (Skarbek et al. 2020). Similar patterns have been observed in other plant community types, both observational and experimental (reviewed by Staab and Schuldt 2020). For example, in oak-pine tree communities, Moreira et al. (2012) showed that increased pine sapling species diversity elevated ant abundance, and Kaitaniemi et al. (2007) found increased abundance of predatory ants in mixed relative to monoculture experimental boreal forest stands. Importantly, increases in ant abundance and diversity in some of these studies (e.g. former of the last two and in BEF-China system) are related to increased abundance of sap-feeding insects tended by ants, which we will review ahead. In addition, studies in temperate experimental grasslands have found that higher plant species richness supports more diverse ant communities (Ebeling et al. 2018; Achury et al. 2022), and, in line with these experimental systems, observational studies such as that by Grevé et al. (2018) reported increased ant abundance and species richness along a gradient of tree diversity in managed temperate forests.

Additionally, there is evidence that plant diversity has uneven effects on different ant functional groups which drive shifts in ant species composition. For example, in an observational study in subtropical forests, Staab et al. (2014) found that tree species richness positively influenced the diversity of predatory ants, but not omnivorous ants. This ant group-specific response was attributed to differences in nesting requirements and the environmental conditions associated with plant diversity. Notably, higher tree diversity was linked to lower soil pH, which had a stronger influence on omnivorous ants, many of which nest in the soil (Staab et al. 2014). In contrast, predatory ants in the system typically nest in the leaf litter and are less directly affected by soil acidity (Staab et al. 2014). In addition, in the BEF-China experiment, Staab et al. (2015) found stronger positive effects of tree diversity on the abundance of ants with mutualistic interactions with sap-sucking Hemiptera than on predatory ants not associated with these insect herbivores, again highlighting the role of these ant-herbivore mutualisms in driving plant diversity effects on ant community structure (see also Schuldt et al. 2017). In a follow-up study, Fornoff et al. (2019) found no effect of tree diversity on the composition of Hemiptera-tending ants in the same system.

8.2.2 Effects of Plant Diversity on Ant-associated Interactions

8.2.2.1 Ant-Herbivore Antagonisms

A handful of studies have examined the effects of plant diversity on ant–herbivore antagonistic interactions, specifically via ant deterrence or predation on herbivores (Rezende et al. 2014; Staab et al. 2023; Sparer et al. 2024). Several of these studies support the prediction of enhanced ant-mediated top-down control of herbivores in more diverse plant communities, with a good amount of this work linking these outcomes to specific plant traits such as extrafloral nectar (EFN) mediating ant-plant mutualisms. For example, a study in the BEF-China system by Staab et al. (2017) found that EFN-producing trees were more likely to be present in diverse tree stands and attracted a more species-rich predatory arthropod community, including ants, as well as spiders and predatory beetles and flies. Then, follow-up work by Staab et al. (2023) in that system found that increased frequency of EFN-producing tree species in tree species polycultures boosted ant recruitment, and this resulted in reduced herbivore abundance and damage on both EFN-bearing and non-EFN plants (via spillover from the former), as well as at the whole-stand level. Relatedly, observational work by Rezende et al. (2014) reported that the presence of EFN-bearing trees near coffee farms is linked to increased ant abundance and species richness on coffee plants. Similar findings have been reported in other agroecosystems such as orchards (see Jones et al. 2017).

There is also evidence of plant diversity effects on ant-herbivore interactions via floral resources, mainly from studies in crop systems with companion (non-crop) plants offering floral resources (Gurr et al. 2017; Jackowicz and Sigsgaard 2025). Although ants responses to floral resources are less well understood than those of other natural enemies, nectar and pollen are frequently exploited resources by ants (Hölldobler and Wilson 1990; Lach 2008). In one of the few available examples, a recent observational study conducted across 37 organic zucchini farms found that the presence and diversity of flowering plants within and around these agroecosystems had no effect on predator abundance, including ants, but was positively associated with abundance of an invasive predator, the red imported fire ant, *Solenopsis invicta* (Sparer et al. 2024). This ant, however, apparently did not influence herbivore suppression (Sparer et al. 2024).

Other studies have tested plant diversity effects on ant–herbivore antagonisms without directly linking outcomes to specific plant traits. For instance, Leles et al. (2017) found that ants were responsible for increased predation rates on artificial caterpillars along a gradient of tree diversity in natural tropical forests, suggesting enhanced herbivore suppression. Similarly, in an experiment using saplings of three pine species, Moreira et al. (2012) found that greater pine species richness was associated with increased ant abundance, accompanied by a decline in the abundance of chewing herbivores. Notably, these effects were absent on trees where ants had been experimentally excluded, highlighting the role of ants in mediating plant diversity-dependent top-down control on herbivores (Moreira et al. 2012). Lastly, Kaitaniemi et al. (2007) found that increased abundance of predatory ants led to increased predation of the European pine sawfly (*Neodiprion sertifer*) in mixed relative to monoculture experimental boreal forest stands.

8.2.2.2 Ant-Herbivore Mutualisms

The diet of omnivorous ants includes not only plant-derived foods (as noted above) but also resources obtained from herbivores (Davidson et al. 2003). Notably, ants often engage in mutualistic relationships with honeydew-producing hemipterans such as aphids, scale insects, and mealybugs (Nelson and Mooney 2022), whereby ants protect these hemipterans from predators and parasitoids in exchange for honeydew—a carbohydrate-rich energy source (reviewed by Stadler and Dixon 2005; Styrsky and Eubanks 2007). These mutualisms can in turn influence other herbivore guilds as well as predators and result in net positive, neutral, or negative effects on overall herbivory and plant performance (Mooney 2007; Styrsky and Eubanks 2007; Nelson and Mooney 2022), thereby complicating predictions about plant diversity effects on ant–mediated effects on food web dynamics and top-down control.

An early study by Riihimäki et al. (2005) provided the first solid evidence for plant diversity effects on ant–hemipteran mutualisms. Conducted in a boreal forest diversity experiment in Finland, the authors reported that ants recruited more strongly to birch trees in mixed stands (birch and pine) than in birch monocultures, as ants spilled over from neighbouring pine trees hosted colonies of ant-tended aphids. In turn, this led to reduced survival of the autumnal moth (*Epirrita autumnata*) on birch trees found in mixtures. Subsequent evidence that plant diversity modulates ant–hemipteran mutualisms comes from the BEF-China experiment in a subtropical forest. Staab et al. (2015) found that increasing tree diversity resulted in a higher frequency of trees hosting ant–hemipteran mutualisms, whereas Fornoff et al. (2019) showed that greater tree diversity promoted the diversity of both mutualistic ants and hemipterans. However, neither of these studies addressed the consequences for top-down effects on herbivores (i.e. predation activity and herbivore suppression). Additional work in other systems has nonetheless documented such dynamics. For example, in a study on plant intraspecific diversity effects, Moreira and Mooney (2013) reported that genotypic polycultures of the shrub *Baccharis salicifolia* supported higher abundances of aphids and ants compared to monocultures, but increased ant recruitment did not affect top-down control of herbivory or plant growth. By contrast, in another study, the same authors found that pine seedling diversity increased aphid and ant abundance (Moreira et al. 2012). They further excluded ants across diversity levels and showed that greater ant recruitment enhanced plant growth in seedling mixtures, presumably because ants deterred or preyed upon non-aphid herbivores (Moreira et al. 2012).

8.2.2.3 Ant-Predator Interactions

Plant diversity can also influence interactions between ants and other predators or parasitoids, which may influence food web structure and top-down control (Staab and Schuldt 2020; Abdala-Roberts and Moreira 2025). Negative interactions such as competition or intraguild predation have been shown to weaken with plant diversity, presumably via increased structural and resource complexity which promote spatial and temporal segregation in foraging activity or ant escape from higher-order predators such as lizards and birds (Abdala-Roberts and Moreira 2025). In addition, plant diversity may foster positive interactions through predator complementarity, whereby different predator species or guilds act synergistically to suppress herbivores (reviewed by Snyder 2019, Abdala-Roberts and Moreira 2025). Though scant, there is some evidence suggestive of these dynamics involving ants which we review next.

One of the first manipulative studies is that by Moreira et al. (2012), who found that increased pine seedling species diversity led to higher ant abundance, which in turn negatively affected the abundance of lady beetles and spiders (the specific interactions or mechanism underlying this effect were not investigated). Interestingly, however, there was an overall positive effect on plant growth mediated by ants (they factorially manipulated ant presence across levels of plant diversity) despite these negative ant-predator interactions (Moreira et al. 2012). Similarly, Moreira and Mooney (2013) found that *B. salicifolia* genotypic mixtures supported higher ant abundance compared to monocultures (mediated by an increase in aphids), and while ants reduced parasitoid abundance, the strength of this interaction and thus the magnitude of parasitoid reduction by ants were similar across levels of plant diversity. Regardless, there were no effects of plant diversity on plant growth mediated by ants, suggesting no consequences for top-down control via ants (Moreira and Mooney 2013). In addition, work from the BEF-China experiment also reported findings suggestive of plant diversity-driven negative interactions between ants and other predators. For example, Staab et al. (2015) found that tree species richness promoted a negative correlation between ant and hunting spider abundance, suggestive of antagonistic interactions, though the mechanisms for this pattern were not investigated. In addition, an observational study by Lucatero et al. (2024) in urban gardens found that increasing plant diversity was associated with reduced abundance of ground-dwelling predators, possibly linked to negative interactions with ants. Finally, observational work by Sparer et al. (2024) found that the diversity of companion (non-crop) plants, via floral resources, promoted invasive ants on zucchini crops which interfered with native predators (including native ants), but again, there was no evidence that this influenced top-down control.

8.3 Implications for Ant-Mediated Trophic Control

Our review highlights different bottom-up pathways by which plant diversity boosts ant abundance and diversity, namely, via plant foods (EFN, floral resources), prey, or mutualistic herbivorous partners, as well as due to improved microhabitat conditions (e.g. for nesting). Ant responses likely also involve changes in foraging or defensive behaviour (e.g. patrolling frequency, aggressiveness), though these ant trait-mediated effects remain largely unstudied. Our review also highlights strong impacts of plant diversity on both ant–herbivore and ant–predator interactions, often linked to increases in ant abundance and diversity, leading to enhanced ant-mediated herbivore suppression in both experimental and natural systems and spanning different plant community types, such as temperate (e.g. Kaitaniemi et al. 2007; Moreira et al. 2012; Leles et al. 2017) and subtropical (Staab et al. 2023) forests, as well as agroecosystems (e.g. coffee farms; Rezende et al. 2014). There was, however, one study where aphid-related increased ant recruitment with plant diversity did not lead to stronger top-down control of non-mutualistic herbivores (Moreira and Mooney 2013).

Interestingly, in several cases, ant-mediated herbivore suppression is observed despite the presence of ant–herbivore mutualisms or negative interactions between ants and other predators that could potentially weaken top-down control (e.g. Moreira et al. 2012; Moreira and Mooney 2013). This suggests that ant effects on pest suppression are strong enough to override both the negative influence of hemipterans and the inhibitory effects of ants on other predatory arthropods beneficial to plants (Abdala-Roberts and Moreira 2025). One explanation for the prevalence of ant top-down effects is that mutualistic herbivores may have relatively weak direct negative impacts on plants, or at least weaker effects compared to non-mutualistic herbivores, as documented in some ant–aphid systems (Styrsky and Eubanks 2007). However, this may not hold for other sap-sucking herbivores tended by ants, such as whiteflies, which can cause more severe plant damage or transmit pathogens (e.g. whiteflies and viruses; Blackman and Eastop 2006). Despite these possibilities, we did not observe weakened herbivore suppression with increased plant diversity. This suggests that, although ant-associated interactions involve trophic complexities and a range of potential outcomes, the net effect of ants is most often neutral to positive for pest suppression and plant performance (Anjos et al. 2022).

At the same time, such interactions are highly variable in both strength and direction and are strongly context-dependent (Chamberlain and Holland 2009). Neutral or negative effects of ants on pest suppression are therefore expected under some circumstances and are likely more common in certain systems. For example, if mutualistic herbivores reach high densities in plant mixtures and are particularly damaging, net negative effects on pest suppression and plant performance may occur (Wielgoss et al. 2014). Similarly, if ants (or other predators) exert weak or no effects on non-mutualistic herbivores, this could lead to neutral diversity effects on top-down control (Schuldt et al. 2017). Negative outcomes may also arise when plant diversity effects are mediated by floral traits that promote antagonistic interactions between ants and parasitoids or pollinators (Jones and Agrawal 2017). Finally, even when ant-mediated suppression of herbivores is achieved in plant mixtures, this does not necessarily translate into benefits for plant growth or reproduction due to trophic cascade attenuation (Schmitz et al. 2000; Mooney et al. 2010). The ultimate outcomes will depend on plant and insect community structure (evenness, composition) as well as the type, frequency, and relative strength of interactions among plants, ants, herbivores, and predators (Abdala-Roberts and Moreira 2025).

8.4 Plant Diversity Effects on Ants Versus Other Predator Groups

There are clearly several mechanisms influencing higher trophic levels that are common to ants and other predator groups, particularly generalist omnivorous predators (Lach et al. 2010; Abdala-Roberts and Moreira 2025). Accordingly, we have thus far (at least implicitly) equated ants with other predaceous arthropods by examining various mechanisms—such as plant food availability, prey abundance, and interactions with both prey and other predators—without making the distinction between ants and other groups of generalist predators. However, each predator group has unique characteristics species- or community-level traits that introduce variability in responses to plant diversity. Ants are not the exception and in some respects could be considered ecological outliers. In this section, we discuss how plant traits mediating the effects of plant diversity (including plant-derived food resources) might affect ants differently—potentially more strongly or in qualitatively different ways—than other predator groups. We also briefly consider ant traits that could help explain the variability and nature of their responses to plant traits mediating diversity effects and the resulting trophic impacts.

Ants are true omnivores, but unlike many other omnivorous predators that opportunistically feed on plant resources, ants engage in much more widespread—nearly ubiquitous—interactions with plants, particularly those mediated by extrafloral nectaries (Hölldobler and Wilson 1990; Rico-Gray and Oliveira 2007; Heil 2015). One obvious expectation, therefore, is that EFN-mediated effects of plant diversity are likely to be stronger and more frequent for ants than for other arthropod predators. In the case of diversity effects mediated by habitat structural complexity—although virtually unstudied for ants—effects may operate differently compared to what has been reported for other predators (Lawton 1983; Andow 1991; Denno et al. 2005). For example, the benefits of reduced intra-guild predation, as reported for other predator groups (Abdala-Roberts and Moreira 2025), may not apply to ants. Ants are often a less preferred prey item for larger predators—except for a few higher-order predators that specialize on ants, such as certain birds—and are also less susceptible to predation due to group foraging, rapid recruitment, and defensive behaviours (Veselý et al. 2021). In addition, ants are highly active foragers with lower dispersal capacity than other predator types (e.g. flying insects), and complex vegetation may increase their foraging energy expenditure (Lucatero et al. 2024), leading to negative effects on ant foraging or abundance. At the same time, however, the effects of structural complexity mediated by plant diversity may positively influence ants via other mechanisms, e.g. increasing the availability of suitable nesting sites in the canopy or leaf litter (Skarbek et al. 2020). Lastly, diversity effects mediated by plant chemical traits—particularly leaf and floral volatiles—may also influence ants in contrasting ways relative to other omnivorous predators that utilize these metabolites. Although there is evidence that ants can use leaf volatiles as foraging cues (Rasmann et al. 2014; Nelson et al. 2020), it remains unclear whether such volatile-mediated signalling is as widespread or ecologically important for ants as it is for other predators (e.g. parasitoids, predatory bugs, or mites). Similarly, floral volatiles can attract ants (e.g. when they act as pollinators; de Vega et al. 2014), but in other cases, they may function as repellents—particularly when ants interfere with pollination (Willmer et al. 2009). Consequently, floral trait-mediated plant diversity effects may lead to more variable recruitment patterns and trophic outcomes for ants compared to those observed in other omnivorous (e.g. pollen-consuming) predators or generalist parasitoids.

In considering these possibilities, an explicit focus on ant traits and population characteristics that shape their responses to plant traits and interactions with herbivores and predators can offer key insights into the underlying mechanisms. In particular, ants' capacity for massive recruitment to food resources or prey, their highly aggregated foraging behaviour and spatial distributions, and their aggressive interactions are largely unique and highly influential. These traits often result in rapid and impactful—but spatially localized—effects, leading to landscape-level spatio-temporal mosaics in species interactions and trophic control (Stüber et al. 2021). This pattern may be especially pronounced in smaller ant species with limited mobility and somewhat less so—though still relevant—for larger, more mobile predatory species. Comparative work examining responses among ant functional groups with contrasting foraging strategies (e.g. solitary vs. group foraging), nesting requirements, mobility, and dietary composition (e.g. relative dependence on plant-derived resources versus animal prey) is especially needed. In this sense, predictions and comparisons of plant diversity effects on ants versus other predator groups can be extended to studies of different ant functional groups.

8.5 Future Work

We argue that the study of ant responses to plant diversity is a missing link that can deepen and broaden our understanding of bottom-up plant controls on community structure and trophic dynamics. Future research should explicitly focus on the plant traits or habitat features that drive plant diversity effects on ants as well as on the ant functional traits that shape their responses to plant diversity and mediate interactions with plants, herbivores, and predators. While our review categorizes different types of ant responses—ranging from community-level outcomes (e.g. composition, diversity) to various ant-associated interactions—these responses are deeply interconnected. Ant community-level patterns both influence and are influenced by their interactions, and different types of ant-associated interactions often exhibit overlapping tri-trophic dynamics (e.g. ant-plant or ant-hemipteran mutualisms that affect other herbivores and ant-predator relationships). Moving forward, synthetic and cross-scale studies will be crucial for developing frameworks and testing theories about how plant diversity shapes ants and their interaction networks. This includes examining effects at the level of individual ant populations and focal interactions (e.g. effects of plant diversity on ant functional responses; Abdala-Roberts and Mooney 2014), as well as broader scales such as ant communities, food web modules, or entire food webs (e.g. changes in ant-associated network structure and related top-down control; Vandermeer et al. 2024). Finally, incorporating spatiotemporal dynamics within and across these levels will be essential to understand the stability and spatial structuring of ant-mediated plant diversity effects.

References

Abdala-Roberts L, Mooney KA (2014) Ecological and evolutionary consequences of plant genotype diversity in a tri-trophic system. Ecology 95:2879–2893

Abdala-Roberts L, Moreira X (2024) Effects of phytochemical diversity on multitrophic interactions. Curr Opin Insect Sci 64:101228

Abdala-Roberts L, Moreira X (2025) Effects of plant diversity on predator-predator interactions in terrestrial ecosystems. In: Gontijo L, Wajnberg E (eds) The ecology of plant diversity: fundamentals and applications. Springer

Abdala-Roberts L, González-Moreno A, Mooney KA, Moreira X, González-Hernández A, Parra-Tabla V (2016) Effects of tree species and genotypic diversity on leafminer-parasitoid interactions. Agric For Entomol 18:43–51

Achury R, Clement L, Ebeling A, Meyer S, Voigt W, Weisser WW (2022) Plant diversity and functional identity alter ant occurrence and activity in experimental grasslands. Ecosphere 13:e4252

Andow DA (1991) Vegetational diversity and arthropod population response. Annu Rev Entomol 36:561–586

Anjos DV, Tena A, Viana-Junior AB, Carvalho RL, Torezan-Silingardi HM, Del-Claro K, Perfecto I (2022) The effects of ants on pest control: a meta-analysis. Proc R Soc B 289:20221316

Barnes AD, Scherber C, Brose U, Borer ET, Ebeling A, Gauzens B, Giling DP, Hines J, Isbell F, Ristok C, Tilman D, Weisser WW, Eisenhauer N (2020) Biodiversity enhances the multitrophic control of arthropod herbivory. Sci Adv 6:eabb6603

Blackman RL, Eastop VF (2006) Aphids on the world's herbaceous plants and shrubs. Wiley, Chichester

Bustos-Segura C, Poelman EH, Reichelt M, Gershenzon J, Gols R (2017) Intraspecific chemical diversity among neighbouring plants correlates positively with plant size and herbivore load but negatively with herbivore damage. Ecol Lett 20:87–97

Cardinale B, Matulich KL, Hooper DU, Byrnes JE, Duffy E, Gamfeldt L, Balvanera P, O'connor MI, González A (2011) The functional role of producer diversity in ecosystems. Am J Bot 98:572–592

Chamberlain SA, Holland JN (2009) Quantitative synthesis of context dependency in ant–plant protection mutualisms. Ecology 90:2384–2392

Davidson DW, Cook SC, Snelling RR, Chua TH (2003) Explaining the abundance of ants in lowland tropical rainforest canopies. Science 300:969–972

de Vega C, Herrera CM, Dötterl S (2014) Floral volatiles play a key role in specialized ant pollination. Perspect Plant Ecol Evol Syst 16:32–42

Denno RF, Finke DL, Langellotto GA (2005) Direct and indirect effects of vegetation structure and habitat complexity on predator-prey and predator-predator interactions. In: Barbosa P, Castellanos I (eds) Ecology of predator-prey interactions. Oxford University Press, New York, pp 211–239

Duffy JE, Cardinale BJ, France KE, McIntyre PB, Thébault E, Loreau M (2007) The functional role of biodiversity in ecosystems: incorporating trophic complexity. Ecol Lett 10:522–538
Ebeling A, Hines J, Hertzog LR, Lange M, Meyer ST, Simons NK, Weisser WW (2018) Plant diversity effects on arthropods and arthropod-dependent ecosystem functions in a biodiversity experiment. Basic Appl Ecol 26:50–63
Ezquivel-Gómez L, Abdala-Roberts L, Pinkus-Rendón M, Parra-Tabla V (2016) Effects of tree species diversity on a community of weaver spiders in a tropical forest plantation. Biotropica
Finke DL, Denno RF (2002) Intraguild predation diminished in complex structured vegetation: implications for prey suppression. Ecology 83:643–652
Fornoff F, Klein AM, Blüthgen N, Staab M (2019) Tree diversity increases robustness of multitrophic interactions. Proc R Soc B 286:20182399
Grevé ME, Hager J, Weisser WW, Schall P, Gossner MM, Feldhaar H (2018) Effect of forest management on temperate ant communities. Ecosphere 9:e02303
Gurr GM, Wratten SD, Landis DA, You M (2017) Habitat management to suppress pest populations: progress and prospects. Annu Rev Entomol 62:91–109
Haddad NM, Crutsinger GM, Gross K, Haarstad J, Knops JMH, Tilman D (2009) Plant species loss decreases arthropod diversity and shifts trophic structure. Ecol Lett 12:1029–1039
Heil M (2015) Extrafloral nectar at the plant-insect interface: a spotlight on chemical ecology, phenotypic plasticity, and food webs. Annu Rev Entomol 60:213–232
Heil M, Mckey D (2003) Protective ant-plant interactions as model systems in ecological and evolutionary research. Annu Rev Ecol Evol Syst 34:425–453
Hertzog LR, Ebeling A, Weisser WW, Meyer ST (2017) Plant diversity increases predation by ground-dwelling invertebrate predators. Ecosphere 8:e01990
Herz A, Heitland W (2005) Species diversity and niche separation of cocoon parasitoids in different forest types with endemic populations of their host, the common pine sawfly *Diprion pini* (hymenoptera: Diprionidae). Eur J Entomol 102:217–224
Hölldobler B, Wilson EO (1990) The ants. Harvard University Press
Holmes KD, Blubaugh CK (2023) A guide to 23 global syntheses of plant diversity effects: unpacking consensus and incongruence across trophic levels. Q Rev Biol 98:121–148
Hooper DU, Chapin FS, Ewel JJ, Hector A, Inchausti P, Lavorel S, Lawton JH, Lodge DM, Loreau M, Naeem S, Schmid B, Setälä H, Symstad AJ, Vandermeer J, Wardle DA (2005) Effects of biodiversity on ecosystem functioning: a consensus of current knowledge. Ecol Monogr 75:3–35
Jackowicz N, Sigsgaard L (2025) Highly diverse flower strips promote natural enemies more in annual field crops: a review and meta-analysis. Agric Ecosyst Environ 381:109412
Jactel H, Moreira X, Castagneyrol B (2021) Tree diversity and forest resistance to insect pests: patterns, mechanisms and prospects. Annu Rev Entomol 66:277–296
Jones PL, Agrawal AA (2017) Learning in insect pollinators and herbivores. Annu Rev Entomol 62:53–71
Jones MI, Koptur S, von Wettberg JE (2017) The use of extrafloral nectar in pest management: overcoming context dependence. J Appl Ecol 54:489–499
Kaitaniemi P, Riihimäki J, Koricheva J, Vehviläinen H (2007) Experimental evidence for associational resistance against the European pine sawfly in mixed tree stands. Silva Fennica 41:259–268
Lach L (2008) Argentine ants displace floral arthropods in a biodiversity hotspot. Divers Distrib 14:281–290
Lach L, Parr CL, Abbott KL (2010) Ant ecology. Oxford University Press, Oxford
Lawton JH (1983) Plant architecture and the diversity of phytophagous insects. Annu Rev Entomol 28:23–39
Leles B, Xiao X, Pasion BO, Nakamura A, Tomlinson KW (2017) Does plant diversity increase top-down control of herbivorous insects in tropical forest? Oikos 126:1142–1149
Letourneau DK, Jedlicka JA, Bothwell SG, Moreno CR (2009) Effects of natural enemy biodiversity on the suppression of arthropod herbivores in terrestrial ecosystems. Ann Rev Ecol Evol Syst 40:573–592

Letourneau DK, Armbrecht I, Salguero-Rivera B, Montoya-Lerma J, Jiménez-Carmona E, Constanza-Daza M, Escobar S, Galindo V, Gutiérrez C, Duque-López S, López-Mejía J, Acosta-Rangel AM, Herrera-Rangel J, Rivera L, Saavedra CA, Torres AM, Reyes-Trujillo A (2011) Does plant diversity benefit agroecosystems? A synthetic review. Ecol Appl 21:9–21

Li Y, Schmid B, Schuldt A, Li S, Wang M-Q, Fornoff F, Staab M, Guo P-F, Anttonen P, Chesters D, Bruelheide H, Zhu C-D, Ma K, Liu X (2023) Multitrophic arthropod diversity mediates tree diversity effects on primary productivity. Nat Ecol Evol 7:832–840

Lucatero A, Jha S, Philpott SM (2024) Local habitat complexity and its effects on herbivores and predators in urban agroecosystems. Insects 15:41

Mooney KA (2007) Tritrophic effects of birds and ants on a canopy food web, tree growth, and phytochemistry. Ecology 88:2005–2014

Mooney KA, Gruner DS, Barber NA, Van Bael SA, Philpott SM, Greenbrerg R (2010) Interactions among predators and the cascading effects of vertebrate insectivores on arthropod communities and plants. Proc Natl Acad Sci USA 107:7335–7340

Moreira X, Mooney KA (2013) Influence of plant genetic diversity on interactions between higher trophic levels. Biol Lett 9:20130133

Moreira X, Mooney KA, Zas R, Sampedro L (2012) Bottom-up effects of host-plant species diversity and top-down effects of ants interactively increase plant performance. Proc R Soc B 279:4464–4472

Moreira X, Abdala-Roberts L, Rasmann S, Castagneyrol B, Mooney KA (2016) Plant diversity effects on insect herbivores and their natural enemies: current thinking, recent findings, and future directions. Curr Opin Insect Sci 14:1–7

Nell CS, Mooney KA (2019) Plant structural complexity mediates trade-off in direct and indirect plant defense by birds. Ecology 100:e02853

Nell CS, Abdala-Roberts L, Parra-Tabla V, Mooney KA (2018) Tropical tree diversity mediates foraging and predatory effects of insectivorous birds. Proc R Soc B 285:20181842

Nelson AS, Mooney KA (2022) The evolution and ecology of interactions between ants and honeydew-producing hemipteran insects. Annu Rev Ecol Evol Syst 53:379–402

Nelson AS, Zapata GD, Sentner KT, Mooney KA (2020) Are ants botanists? Ant associative learning of plant chemicals mediates foraging for carbohydrates. Ecol Entomol 45:251–258

Rasmann S, Buri A, Gallot-Lavalee M, Joaquim J, Purcell J, Pellissier L (2014) Differential allocation and deployment of direct and indirect defences of *Vicia sepium* along elevation gradients. J Ecol 102:930–938

Rezende MQ, Venzon M, Perez AL, Cardoso IM, Janssen A (2014) Extrafloral nectaries of associated trees can enhance natural pest control. Agric Ecosyst Environ 188:198–203

Rico-Gray V, Oliveira PS (2007) The ecology and evolution of ant-plant interactions. The University of Chicago Press, Chicago

Riihimäki J, Kaitaniemi P, Koricheva J, Vehviläinen H (2005) Testing the enemy hypothesis in forest stands: the important role of tree species composition. Oecologia 142:90–97

Santos de Araújo W, Daud RD (2018) Investigating effects of host-plant diversity on Brazilian mite richness in natural ecosystems. Syst Appl Acarol 23:1598

Scherber C, Eisenhauer N, Weisser WW, Schmid B, Voigt W, Fischer M, Schulze ED, Roscher C, Weigelt A, Allan E, Bessler H, Bonkowski M, Buchmann N, Buscot F, Clement LW, Ebeling A, Engels C, Halle S, Kertscher I, Klein AM, Koller R, Konig S, Kowalski E, Kummer V, Kuu A, Lange M, Lauterbach D, Middelhoff C, Migunova VD, Milcu A, Muller R, Partsch S, Petermann JS, Renker C, Rottstock T, Sabais A, Scheu S, Schumacher J, Temperton VM, Tscharntke T (2010) Bottom-up effects of plant diversity on multitrophic interactions in a biodiversity experiment. Nature 468:553–556

Schmitz OJ, Hamback PA, Beckerman AP (2000) Trophic cascades in terrestrial systems: a review of the effects of carnivore removal on plants. Am Nat 155:141–153

Schuldt A, Staab M (2015) Tree species richness strengthens relationships between ants and the functional composition of spider assemblages in a highly diverse forest. Biotropica 47:339–346

Schuldt A, Fornoff F, Bruelheide H, Klein AM, Staab M (2017) Tree species richness attenuates the positive relationship between mutualistic ant-hemipteran interactions and leaf chewer herbivory. Proc R Soc B 284:20171489

Schuldt A, Ebeling A, Kunz M, Staab M, Guimarães-Steinicke C, Bachmann D, Buchmann N, Durka W, Fichtner A, Fornoff F, Härdtle W, Hertzog L, Klein AM, Roscher C, Schaller J, von Oheimb G, Weigelt A, Weisser W, Wirth C, Zhang J, Bruelheide H, Eisenhauer N (2019) Multiple plant diversity components drive consumer communities across ecosystems. Nat Commun 10:1460

Skarbek CJ, Noack M, Bruelheide H, Härdtle W, von Oheimb G, Scholten T, Seitz S, Staab M (2020) A tale of scale: community but not neighborhood tree diversity increases leaf litter ant diversity. J Anim Ecol 89:299–308

Snyder WE (2019) Give predators a complement: conserving natural enemy biodiversity to improve biocontrol. Biol Control 135:73–82

Sparer AE, Madden MK, Blubaugh CK (2024) Flower diversity on organic farms increases invasive ants rather than non-invasive natural enemies. Agric For Entomol 26:159–170

Staab M, Schuldt A (2020) The influence of tree diversity on natural enemies – a review of the "enemies" hypothesis in forests. Curr For Rep 6:243–259

Staab M, Schuldt A, Assmann T, Klein A-M (2014) Tree diversity promotes predator but not omnivore ants in a subtropical Chinese forest: tree diversity promotes predator ants. Ecol Entomol 39:637–647

Staab M, Blüthgen N, Klein A-M (2015) Tree diversity alters the structure of a tri-trophic network in a biodiversity experiment. Oikos 124:827–834

Staab M, Methorst J, Peters J, Blüthgen N, Klein A-M (2017) Tree diversity and nectar composition affect arthropod visitors on extrafloral nectaries in a diversity experiment. J Plant Ecol 10:201–212

Staab M, Liu X, Assmann T, Bruelheide H, Buscot F, Durka W, Erfmeier A, Klein A-M, Ma K, Michalski S, Wubet T, Schmid B, Schuldt A (2021) Tree phylogenetic diversity structures multitrophic communities. Funct Ecol 35:521–534

Staab M, Pietsch S, Yan H, Blüthgen N, Cheng A, Li Y, Zhang N, Ma K, Liu X (2023) Dear neighbor: trees with extrafloral nectaries facilitate defense and growth of adjacent undefended trees. Ecology 104:e4057

Stadler B, Dixon AFG (2005) Ecology and evolution of aphid-ant interactions. Annu Rev Ecol Evol Syst 36:345–372

Stemmelen A, Paquette A, Benot M-L, Kadiri Y, Jactel H, Castagneyrol B (2022) Insect herbivory on urban trees: complementary effects of tree neighbours and predation. Peer Commun J 2:e22

Stüber M, Tack AJM, Zewdie B, Mendesil E, Shimales T, Ayalew B, Nemomissa S, Sjögren J, Vesterinen E, Wezel A, Hylander K (2021) Multi-scale mosaics in top-down pest control by ants from natural coffee forests to plantations. Ecology 102:e03376

Styrsky JD, Eubanks MD (2007) Ecological consequences of interactions between ants and honeydew-producing insects. Proc R Soc B Biol Sci 274:151–164

Tilman D, Isbell F, Cowles JM (2014) Biodiversity and ecosystem functioning. Annu Rev Ecol Evol Syst 45:471–493

Vandermeer J, Hajian-Forooshani Z, Rivera-Salinas IS, Perfecto I (2024) Pest control in coffee: a tri-trophic comparison between a mainland and an Island agroecosystem. In: Moreira X, Abdala-Roberts L (eds) Ecology and evolution of plant-herbivore interactions on islands. Ecological studies 249. Springer Nature, Cham

Veselý P, Dobrovodský J, Fuchs R (2021) Predation by avian predators may have initiated the evolution of myrmecomorph spiders. Sci Rep 11:17266

Wan N-F, Zheng X-R, Fu L-W, Kiær LP, Zhang Z, Chaplin-Kramer R, Dainese M, Tan J, Qiu S-Y, Hu Y-Q, Tian W-D, Nie M, Ju R-T, Deng J-Y, Jiang J-X, Cai Y-M, Li B (2020) Global synthesis of effects of plant species diversity on trophic groups and interactions. Nat Plants 6:503–510

Wielgoss A, Tscharntke T, Rumede A, Fiala B, Seidel H, Shahabuddin S, Clough Y (2014) Interaction complexity matters: disentangling services and disservices of ant communities driving yield in tropical agroecosystems. Proc R Soc B 281:20132144

Willmer PG, Nuttman CV, Raine NE, Stone GN, Pattrick JG, Henson K, Stillman P, McIlroy L, Potts SG, Knudsen JT (2009) Floral volatiles controlling ant behaviour. Funct Ecol 23:888–900

Zou Y, Sang W, Bai F, Axmacher JC (2013) Relationships between plant diversity and the abundance and α-diversity of predatory ground beetles (Coleoptera: Carabidae) in a mature Asian temperate forest ecosystem. PLoS One 8:e82792

The Effects of Tree Diversity on Insectivorous Vertebrate Communities and Predatory Activity

9

Carla Vázquez-González, Xoaquín Moreira, and Luis Abdala-Roberts

Abstract

Plant diversity effects on communities and ecosystems are inherently linked to multi-trophic interactions. In this chapter, we review the effects of tree diversity on insectivorous birds and bats, two key vertebrate predators with outsized impacts on forest food webs and ecosystem dynamics. We examine the mechanisms through which tree diversity influences these predator communities and their foraging activity, review empirical evidence, and discuss the implications of these effects for trophic control. We conclude by pointing to important avenues for future research. Overall, current evidence indicates consistent positive effects of tree diversity on the abundance and diversity of insectivorous birds and bats across forest types, whereas effects on predatory activity (e.g. predation rates) show similar but more variable trends. In turn, cascading feedbacks on herbivory and plant performance remain largely unstudied, representing a fundamental gap in research. Accordingly, we outline several research directions for developing a predictive framework for tree diversity effects on insectivorous vertebrate-driven trophic cascades. These include explicit attention to tree diversity-associated features such as structural heterogeneity, prey and resource heterogeneity, and microclimatic buffering, as well as consideration of predator traits, prey identity, tree species composition, and spatiotemporal scaling dynamics. Overall, this chapter underscores the importance of promoting diverse tree assemblages and their associated predator communities and functions while calling for mechanistic studies that can inform tree diversity-based strategies to enhance forest resilience.

C. Vázquez-González (✉) · X. Moreira
Misión Biológica de Galicia (MBG-CSIC), Pontevedra, Spain
e-mail: cvazquez@mbg.csic.es

L. Abdala-Roberts
Universidad Nacional Autónoma de México, Escuela Nacional de Estudios Superiores-Unidad Mérida, Ucú, Yucatán, Mexico

X. Moreira et al. (eds.), *Effects of Plant Diversity on Plant-Herbivore Interactions*, Ecological Studies 255, https://doi.org/10.1007/978-3-032-26461-9_9

9.1 Introduction

Plant diversity is a well-established driver of ecosystem processes (Gamfeldt et al. 2013; Tilman et al. 2014; van Der Plas et al. 2016; Ratcliffe et al. 2017; Huang et al. 2018), particularly through its influence on multitrophic interactions among insect herbivores and their natural enemies (Moreira et al. 2016; Albert et al. 2022). This tri-trophic paradigm traces back to Elton's foundational ideas on diversity and ecosystem stability (Elton 1958) and Root's (1973) enemies' hypothesis, which posits that diverse plant communities promote predator abundance and diversity, which in turn suppress and stabilize prey population dynamics. Empirical support initially came mainly from herbaceous systems, especially agroecosystems (Langellotto and Denno 2004; Haddad et al. 2009; Letourneau et al. 2011). More recently, forest studies also show that higher tree diversity can enhance the abundance and diversity of natural enemies and, in turn, increase predation pressure on herbivores in mixed stands (Staab and Schuldt 2020; Stemmelen et al. 2022a). Accordingly, these feedbacks could explain reduced insect herbivore damage and increased primary productivity in restored or experimental (i.e. planted) and naturally occurring mixed forests (Grossman et al. 2019; Poeydebat et al. 2021; Jactel et al. 2021; Li et al. 2023). Yet, despite growing interest, evidence from tree communities remains somewhat limited with respect to effects on predator community structure and foraging behaviour (Stemmelen et al. 2022a). Further, research explicitly linking tree diversity to predator top-down effects on herbivory and plant productivity (i.e. trophic cascades) remains particularly scarce. Such paucity of studies constrains our understanding of the multitrophic processes through which tree diversity regulates food dynamics that support forest biodiversity and ecosystem functioning.

The effects of plant diversity on invertebrate predators, particularly predatory arthropods, are well documented (Letourneau et al. 2011; Abdala-Roberts and Moreira 2026; Moreira and Abdala-Roberts 2026) and include evidence from studies in forests (Stemmelen et al. 2022a, for ants see Moreira and Abdala-Roberts 2026). In contrast, much less is known about vertebrate predator responses to tree diversity, particularly in forests. Available studies have focused on insectivorous birds and, to some extent, also bats (Van Bael et al. 2003, 2008; Kalka et al. 2008; Mooney et al. 2010; Maas et al. 2016), whereas reptiles and small mammals remain virtually unstudied (Kollberg et al. 2014; Pianka et al. 2017). A recent meta-analysis underscores this gap: Of 245 case studies, only ~23% examined bird responses to plant diversity, ~7% focused on bats, and none addressed reptiles or non-volant predatory mammals (Stemmelen et al. 2022a). Further, evidence linking tree diversity to vertebrate predation pressure is relatively rare, especially for experimental studies consisting of local-scale tree diversity manipulations, with only 11 studies conducted to date (most during the last decade, included in this review), of which the majority report positive effects ($n = 7$), a few find null outcomes ($n = 3$), and 1 a negative effect ($n = 1$) (Table 9.1). The need to fill this gap is clear: Forests host 78% of all bird species, with roughly one third strictly dependent on them, and over 80% of all bat species, of which roughly one quarter are forest specialists, and with both taxonomic groups being heavily dominated by insectivorous species

Table 9.1 Summary of empirical studies testing the effects of tree diversity on insectivorous vertebrates (birds and bats), focusing on predator abundance, diversity, and predation activity. For each study, we indicate the type of response variable, the effect of tree diversity (positive, null, or negative), the study type (experimental or observational), and the forest biome (tropical, subtropical, temperate, or boreal). Collectively, these studies illustrate the variety of approaches and outcomes and highlight the context dependency of tree diversity effects on vertebrate-mediated top-down control

Study	Response variable	Effect of tree diversity	Study type	Forest type
Hobson and Bayne (2000)	Bird abundance and richness	Positive	Observational	Boreal
Díaz (2006)	Bird richness	Positive	Observational	Temperate (Mediterranean)
Gil-Tena et al. (2007)	Bird richness	Positive	Observational	Temperate (Mediterranean)
Matlock and Edwards (2006)	Bird abundance and richness	Positive	Observational	Tropical
Kissling et al. (2007)	Bird richness (frugivores)	Positive	Observational	Tropical (sub-Saharan Africa)
Charbonnier et al. (2016)	Bat abundance and richness	Positive	Observational	Temperate
Buchholz et al. (2020)	Bat richness and activity	Positive	Observational	Temperate
Ampoorter et al. (2020)	Bird and bat diversity	Positive (birds); null (bats)	Observational	Temperate (continental Europe)
Penone et al. (2019)	Bird and bat abundance and richness	Null	Observational	Temperate
Nell et al. (2018)	Bird abundance, richness and functional and phylogenetic diversity	Positive	Experimental	Tropical
May-Uc et al. (2020)	Bird abundance, richness and functional and phylogenetic diversity	Positive	Experimental	Tropical
Vázquez-González et al. (2024b)	Bird abundance, richness and functional and phylogenetic diversity	Positive	Experimental	Tropical
Interian-Aguiñaga et al. (2022)	Bird abundance and diversity	Null	Experimental	Tropical

(continued)

Table 9.1 (continued)

Study	Response variable	Effect of tree diversity	Study type	Forest type
Coyne et al. (2025)	Bird abundance	Positive	Experimental	Temperate
Nell et al. (2018)	Bird predation (attack rates)	Positive	Experimental	Tropical
May-Uc et al. (2020)	Bird predation (attack rates)	Positive	Experimental	Tropical
Vázquez-González et al. (2024b)	Bird predation (attack rates)	Positive	Experimental	Tropical
Muiruri et al. (2016)	Bird predation (attack rates)	Positive at tree-neighbour scale	Experimental	Boreal
Coyne et al. (2025)	Bird predation (sentinel prey and exclusions)	Mixed (higher attack in monocultures; but stronger exclusion effects in mixtures)	Experimental	Temperate
Vázquez-González et al. (2024a)	Bird predation (attack rates)	Positive (global synthesis TreeDivNet)	Experimental	Global (multi-biome)
Stemmelen et al. (2022b)	Bird predation (attack rates)	Positive	Observational	Urban/temperate
Schillé et al. (2025)	Bird predation (attack rates)	Positive	Observational	Urban (multi-city)
Castagneyrol et al. (2017)	Bird predation (attack rates)	Null	Experimental	Temperate
Yang et al. (2018)	Bird predation (attack rates)	Null	Experimental	Temperate
Interian-Aguiñaga et al. (2022)	Bird predation (attack rates)	Null	Experimental	Tropical
Nixon and Roland (2012)	Bird predation (attack rates)	Negative	Observational	Boreal
Anttonen et al. (2023)	Bird predation (attack rates)	Variable/ context-dependent	Experimental	Subtropical

(Sekercioglu 2012; Maas et al. 2016). Addressing insectivorous bird and bat responses to tree diversity is therefore key for explaining forest pest regulation and forest resilience, a challenge that requires experimental approaches that manipulate tree diversity to robustly assess these trophic feedbacks and their underlying mechanisms.

Tree diversity may affect insectivorous vertebrate communities and their foraging pressure through several mechanisms. Often-cited ones include greater structural complexity in mixed forests, which creates more foraging niches and nesting substrates (MacArthur and MacArthur 1961; Abdala-Roberts and Moreira 2026; Moreira and Abdala-Roberts 2026), as well as richer and functionally more heterogeneous prey communities and complementary resources (for omnivorous predators) such as fruits, nectar, and cavities (Holmes and Schultz 1988; Russell 1989). At the same time, diverse forests also favour microclimatic buffering, which benefits predators sensitive to temperature and humidity extremes (Chen et al. 1999; Barton and Schmitz 2009; De Frenne et al. 2019). Despite incipient evidence for some of these causal pathways for some plant community types and agroecosystems (Moreira and Abdala-Roberts 2026), explicit mechanistic tests in forest ecosystems remain rare, leaving considerable uncertainty about the consistency and strength of tree diversity effects on insectivorous vertebrates and their top-down effects in forests.

In this chapter, we synthesize current knowledge on tree diversity effect on insectivorous vertebrate communities across different forest types (Fig. 9.1). We begin by introducing the major insectivorous vertebrate groups that regulate herbivores in forests, focusing primarily on birds and, to a lesser extent (given more limited work), on bats. We then review evidence for how tree diversity influences insectivorous bird and bat communities and their predatory activity, as well as the potential

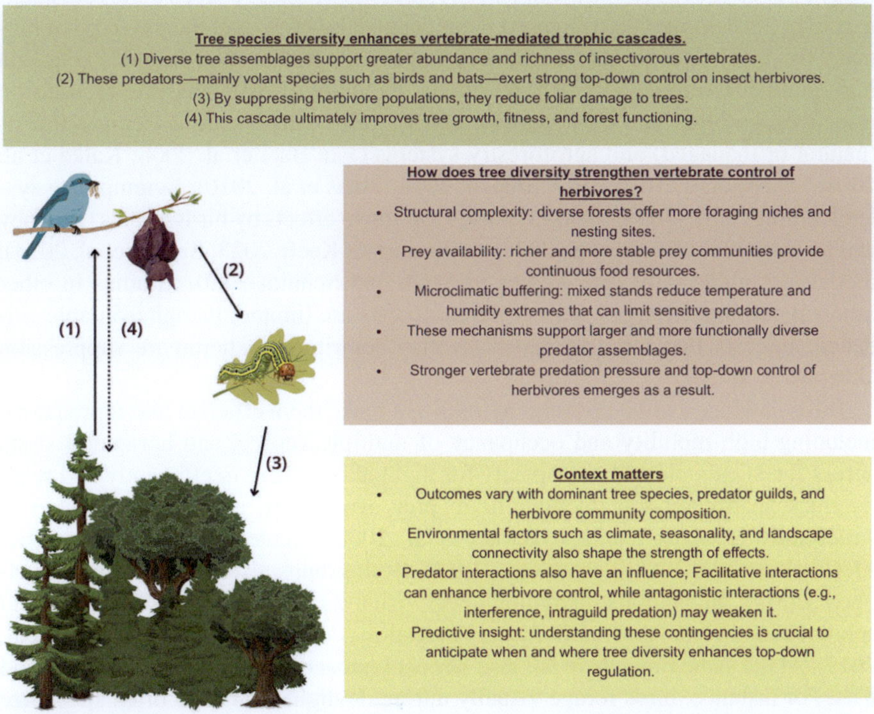

Fig. 9.1 Infographic summarizing the main insights of this chapter: Tree species diversity can strengthen vertebrate-mediated trophic cascades by enhancing predator communities and suppressing herbivory. These effects arise through structural, resource, and microclimatic mechanisms but remain highly context dependent

for bottom-up and top-down feedbacks involving trophic cascades on herbivory and tree productivity. Our review examines different mechanisms through which tree diversity can affect insectivorous vertebrates and also highlights context-dependent outcomes and potential sources behind such variability. Throughout, we cover both experimental studies (i.e. local-scale tree diversity manipulations) and observational studies conducted at different spatial scales (e.g. along natural tree diversity gradients). We end by outlining critical knowledge gaps and proposing research priorities to advance our understanding of the multitrophic pathways linking tree diversity, vertebrate predators, and ecosystem function.

9.2 Insectivorous Vertebrates as a Top-Down Force in Forests

Insectivorous vertebrates play a pivotal role in regulating herbivore pressure, often triggering trophic cascades that enhance plant performance and modulate ecosystem function (Mooney et al. 2010; Maas et al. 2016). A landmark meta-analysis by Mooney et al. (2010) showed that insectivorous vertebrates reduced herbivore abundance by 39%, indirectly decreased plant damage by 40%, and increased plant biomass by 14%, despite also negatively impacting predatory arthropods (e.g. via intra-guild predation). Most of the studies included in their analysis focused on insectivorous birds and bats, particularly in well-studied natural tropical forests (natural or managed) and agroforestry systems (Van Bael et al. 2008; Kalka et al. 2008; Sekercioglu 2012; Perfecto et al. 2014; Maas et al. 2016). In temperate systems, studies have also similarly reported top-down effects by birds or bats (Mooney and Linhart 2006; Böhm et al. 2011; Beilke and O'Keefe 2023; Sivault et al. 2024), though not universally (see studies in Staab and Schuldt 2020). Studies in other arboreal community types such as boreal forests are limited, though available evidence suggests that birds in particular also contribute to herbivore suppression (Zvereva et al. 2020).

Birds and bats exhibit convergent traits that make them efficient insect predators, including high mobility and occupancy of multiple vertical and horizontal strata across forest canopies. Their capacity for flight allows them to efficiently and readily track spatiotemporal fluctuations in prey availability, even in fragmented or structurally complex habitats (Barber et al. 2008; Richter and Cumming 2008; McCracken et al. 2012; Sekercioglu 2012). Both groups also exhibit highly developed sensory adaptations, including acute vision and olfactory sense in birds and echolocation in bats, that facilitate prey detection (Fujioka et al. 2016; Rubene et al. 2019). At the same time, there are also key ecological differences between birds and bats. For instance, birds forage visually during daylight hours and often specialize on large, exposed herbivores such as caterpillars, whereas bats hunt at night using echolocation, often preying on smaller, nocturnal arthropods (Maas et al. 2016). In addition, many insectivorous birds, such as bark-probers or foliage specialists, tend to specialize on herbivorous insects, often targeting specific guilds such as wood-borers or caterpillars (Sherry 1984; Fayt et al. 2005; Whelan et al. 2008; Koenig et al. 2013). Bats tend to be more generalist predators with flexible foraging strategies, enabling them to respond to shifting prey, availability (Kunz et al. 2011), though some functional groups may specialize on specific types of insect prey (e.g. based on echolocation system; Schnitzler et al. 2003). Overall, while dietary generalist species our gilds of bats and birds may deliver more consistent arthropod community-wide suppression, specialists can be critical for limiting particular insect pests, such that jointly, different functional groups and species composition within each is key for predicting the occurrence and strength of trophic cascades by these predator groups.

Habitat characteristics exert strong controls over these insectivorous vertebrates. For instance, bird diversity and associated predation pressure often increase with greater vertical stratification and variation in canopy height, features that increase foraging niche space, nesting sites, and predator refuges (MacArthur and MacArthur 1961; Bradbury et al. 2005; Jankowski et al. 2013; Bereczki et al. 2014; Brady and Farrer 2024). Insectivorous birds also benefit from greater availability and temporal stability of prey resources such as caterpillars and leaf miners (Karr 1976; Holmes and Schultz 1988). Similarly, bats rely on roosting structures such as cavities, exfoliating bark, and crevices and benefit from moderate canopy openness, which enhances manoeuvrability and facilitates prey detection via echolocation (Schnitzler and Kalko 2001, Loeb and O'keefe 2006, Jung et al. 2012, Dodd et al. 2012). In some instances, high vegetation complexity can interfere with bat or bird foraging (Ampoorter et al. 2020; Anttonen et al. 2023), suggesting context-dependencies as a function of factors such as tree species composition or vegetation type and insectivore composition affecting foraging behaviour and efficiency. Additionally, both predator groups may respond to microclimatic buffering effects of tree diversity via canopy complexity and increased forest cover (e.g. via canopy packing) which reduce temperature extremes and humidity loss, factors that not only affect thermoregulation but also indirectly benefit birds and bats via increased arthropod prey availability (Chen et al. 1999; Betts et al. 2018; Kim et al. 2022). These habitat features, combined with changes in predator traits or composition, will determine predator responses and the outcome of predator-prey interactions and top-down control, and should ideally be jointly considered in mechanistic frameworks linking tree diversity and trophic control.

9.3 Tree Diversity Effects on Insectivorous Vertebrate Communities

Empirical evidence linking tree diversity and insectivorous vertebrates have often come from observational investigations in natural forest fragments, across monoculture vs. mixture plantations, or along forest diversity gradients. In contrast, evidence from experimental studies involving tree diversity manipulations are less common (Staab and Schuldt 2020; Stemmelen et al. 2022a). Recently, however, studies from experimental systems within *TreeDivNet* (www.treedivnet.ugent.be;

Paquette et al. 2018), for example, a global network of tree diversity experiments, have started to yield robust evidence for effects on predator responses and top-down control.

In this section, we synthesize current evidence on tree diversity effects on insectivorous birds and bats from local-scale experimental studies (e.g. from the *TreeDivNet* platform), as well as from observational studies in both temperate and tropical systems, focusing on predator community- or population-level responses, as well as predatory activity (foraging behaviour or predation rates), and lastly discuss the implications these effects have for trophic cascades (Table 9.1).

9.3.1 Effects on Bird and Bat Abundance and Diversity

Within the realm of observational studies, comparisons across monoculture and polyculture plantations have proven useful. In Canadian boreal forests, for example, Hobson and Bayne (2000) found that mixed plantations of aspen, pine, and spruce supported higher bird abundance and species richness than pure stands, with some insectivores (e.g. Tennessee Warbler, Swainson's Thrush) strongly associated with tree species mixtures. Likewise, Charbonnier et al. (2016) found that deciduous tree cover (e.g. oak), often associated with higher tree diversity, significantly increased bat activity and species richness compared to monospecific pine stands. Similarly, long-term acoustic monitoring in Germany showed that although monospecific plantations can support diverse bat assemblages, mixed stands generally offer more favourable conditions (Buchholz et al. 2020). In addition, supporting these results, a recent review showed that mixed-species tree plantations overall support higher vertebrate biodiversity than monocultures, with gains being especially strong and consistent for birds (Kremer et al. 2025).

Observational studies have also employed natural forest gradients or forest type comparisons, such as in Mediterranean ecosystems of Central Spain where Díaz (2006) showed that mixed oak–pine forests hosted more bird species than pure pine or oak stands, while in Catalonia, bird richness increased significantly with tree-species diversity, especially for forest specialists (Gil-Tena et al. 2007). Likewise, across six major forest regions in Europe, a study showed that tree species richness positively correlated with the diversity and abundance of insectivorous birds but not bats (Ampoorter et al. 2020). In addition, Penone et al. (2019) found bird and bat abundance and species richness were unrelated to tree species richness across sites varying in tree diversity in a temperate forest in Germany. Finally, in tropical forests, observational work reported positive associations between tree diversity and both predator groups. For instance, in Costa Rican tropical forest fragments, bird richness and abundance increased significantly with tree richness (Matlock Jr and Edwards 2006), while (Kissling et al. 2007) found a positive relationship between *Ficus* richness and frugivorous bird species richness across sub-Saharan Africa.

In addition, although relatively less common, recent studies involving local-scale experimental manipulations of tree diversity (e.g. systems part of *TreeDivNet*) have yielded causal evidence spanning tropical to temperate forests, mainly for insectivorous birds. For example, in the UADY-UNAM tree-diversity experiment in subtropical Yucatán, Mexico, studies have shown that tree polycultures increase insectivorous bird abundance, richness, as well as functional and phylogenetic diversity (Nell et al. 2018; May-Uc et al. 2020; Vázquez-González et al. 2024b), and that these bird community features were more stable across seasons in tree mixtures relative to monoculture (May-Uc et al. 2020). Similarly, Coyne et al. (2025) found that bird abundance was higher in tree polycultures than in monoculture stands at a temperate deciduous forest experimental site. Together, these experimental studies highlight both the potential and the variability of tree diversity effects on insectivorous bird communities, underscoring the need for long-term and cross-site comparisons.

9.3.2 Effects on Bird and Bat Predation Rate and Foraging Activity

Studies on the effects of tree diversity on insectivorous bird and bat predatory activity are limited, but a few examples exist, mostly from the last decade. In the majority of cases, predation has been quantified with artificial-prey models (e.g. plasticine caterpillars) and, to a lesser extent, with bird or bat exclusions. Because attacks by birds on artificial prey are easy to identify (Low et al. 2014), most of this work has reported tree diversity effects on attack rates by this predator group. Overall, results indicate that tree diversity promotes bird predation rates (i.e. attack attempts). For instance, in the UADY-UNAM *TreeDivNet* system in Yucatán, polycultures not only supported richer bird communities (as discussed above) but also exhibited higher avian attack rates on plasticine caterpillars, providing causal links between tree diversity, bird communities, and predation pressure (Nell et al. 2018, May-Uc et al. 2020, Vázquez-González et al. 2024b). Similarly, in a boreal tree diversity experiment in Satakunta, Finland, also part of *TreeDivNet*, bird attack rates were higher on focal trees surrounded by more diverse tree neighbourhoods (Muiruri et al. 2016). In addition, a recent study by Coyne et al. (2025) used both sentinel prey (mealworms) and bird exclusions in an experimental deciduous temperate forest system. They reported higher attack rates in less diverse tree plots but also found stronger effects of bird exclusion in tree mixture which suggested stronger top-down control with tree diversity (Coyne et al. 2025). A global synthesis across *TreeDivNet* sites supports these findings by showing that a tree diversity gradient spanning the analysed systems correlated positively with bird attack rates on dummy caterpillars (Vázquez-González et al. 2024a). Lastly, observational studies in urban forests in Montreal reported that higher tree diversity was associated with increased attack rates on artificial caterpillars and reduced foliar herbivory (Stemmelen et al. 2022b), and a larger dataset similarly recorded higher avian predation attempts with greater tree diversity across rural and urban sites, independently of bird diversity (Schillé et al. 2025).

There are a few studies that have report no effect of tree diversity on predatory activity by insectivorous birds, several of which include local-scale tree diversity manipulations (Castagneyrol et al. 2017; Yang et al. 2018; Interian-Aguiñaga et al. 2022), and there is at least one observational study comparing aspen monoculture and mixed plantations that reported a negative effect (Nixon and Roland 2012). Accordingly, a recent global meta-analysis showed that while tree diversity consistently promoted the abundance and richness of insectivorous birds and bats (see studies above), its effects on predation activity were more variable (Stemmelen et al. 2022a). Several studies highlight this variability, with outcomes contingent on spatial scale (Muiruri et al. 2016), predator functional traits (Vázquez-González et al. 2024b), and intra- and interannual temporal variation (Anttonen et al. 2023; Vázquez-González et al. 2024b). Together, these findings emphasize that while tree diversity often enhances insectivorous vertebrate predation, its effects are not universal and may depend on the ecological context (see Sect. 9.4).

9.3.3 Implications for Insectivorous Vertebrate-Driven Trophic Cascades

To our knowledge, no study to date has yet reported on tree diversity-driven trophic cascades involving insectivorous vertebrates (i.e., ultimate positive effects on trees via reduced herbivory and increased productivity). There are, however, a few studies suggesting this could be the case which we draw upon to discuss potential implications. In particular, a recent experimental study involving bird exclusions showed that tree diversity boosted bird abundance and foraging, which in turn results in stronger top-down control of insect herbivory in tree mixtures (Coyne et al. 2025). Although effects on tree growth were not tested, similar patterns in other systems also suggest a potential for trophic cascades (Haddad et al. 2009; Letourneau et al. 2011). In addition, a few observational studies suggest context-dependency in the effects of structural complexity, seasonality, and prey distribution influencing predator-prey interactions (e.g. augmented intra-guild predation) in ways that may weaken or even reverse plant diversity-mediated trophic cascades (e.g. Nixon and Roland 2012; Ampoorter et al. 2020). In addition, tree diversity may also expose some bird species to higher predation risk (e.g. Batáry et al. 2014). Finally, non-consumptive (behaviourally mediated) pathways are crucial to assess effects on herbivory and trophic cascades but remain largely untested in tree diversity experiments. For example, insectivorous bats can suppress moth activity or mating via acoustic cues (Schmitz et al. 1997; Preisser et al. 2005; Kalka et al. 2008; Kunz et al. 2011). In sum, evidence to date supports the prediction that tree diversity amplifies insectivorous bird-mediated control of herbivores, with potential but largely untested consequences for plant communities. Formal tests of tree diversity-mediated trophic cascades, as well as their mechanisms (e.g. bird community metrics that drive effects, consumptive vs. non-consumptive effects), including predator exclusions (teasing apart bird vs. bat predation effect; Beilke and O'Keefe 2023), are needed to demonstrate these feedbacks and their effects on forest food webs and ecosystem functioning.

9.4 Mechanisms of Tree Diversity Effects on Insectivorous Vertebrates

Despite evidence that tree diversity shapes vertebrate predator communities and foraging activity, the ecological mechanisms driving these effects remain poorly understood. Insights from classical theory and empirical work can be grouped into three categories of mechanisms: (i) habitat structural complexity, (ii) microclimatic conditions, and (iii) prey or resource heterogeneity. These pathways are not mutually exclusive and, in many cases, likely operate simultaneously, with their relative importance varying within and across time (e.g., seasons or years) or forest community types. We next review the available evidence and, where data are lacking, speculate about these mechanisms by drawing on the broader ecological literature.

9.4.1 Tree Diversity Effects Are Driven by Forest Structural Complexity

Positive effects of tree diversity on insectivorous vertebrates are often attributed to plant chemical and physical or structural traits (see Felix and Koricheva 2006 in this volume). In particular, increased structural complexity in mixed forests is frequently invoked when assessing effects on higher trophic levels. Classic work show that bird diversity increases with foliage height diversity, as structurally complex forests provide more foraging and nesting opportunities (MacArthur and MacArthur 1961). In this context, mixed stands feature more variable crown architectures and greater vertical stratification, increasing overall stand structural complexity (Juchheim et al. 2020). Accordingly, these structural attributes are expected to expand niche space for vertebrate predators, promoting habitat partitioning and reducing competition, thereby boosting predation pressure across herbivore guilds (MacArthur and MacArthur 1961; Schuldt et al. 2019). A study by Schuldt et al. (2019) provided evidence for this pathway: in the BEF-China subtropical forest experiment, tree richness increased structural diversity, which in turn explained higher richness and abundance of predatory birds. Interestingly, a latter study in this system by Anttonen et al. (2023) found that increased structural complexity correlated with reduced detectability of prey by insectivorous birds, suggesting these effects change with time as a function of forest structure or changes in bird functional groups and their associated foraging behaviours or strategies (Whelan 2001). In addition, in the Satakunta boreal system in Finland, predation was enhanced at the tree–neighbour scale but not at the plot scale, suggesting that within-stand local crown variation, rather than stand-level richness, drove predation (Muiruri et al. 2016). In contrast, to our knowledge, no studies have directly examined how tree diversity-mediated structural complexity influences bat foraging and top-down effects, though a broad scale study across forest type in Europe suggested vertical complexity may interfere with foraging by some bat functional groups (Ampoorter et al. 2020). The influence of structural complexity on bats is likely strongly contingent on foraging behaviour as a function of echolocation strategy. Short-range echolocators, adapted to cluttered habitats and often targeting non-volant prey, may benefit more from structurally complex and diverse forests. Long-range echolocators, by contrast, are better suited to open habitats and aerial prey, making them less dependent on tree diversity and architecture. Accordingly, addressing insectivorous vertebrate responses to tree diversity as a function of structural complexity must consider species- and functional group-specific traits.

9.4.2 Diversity Effects via Buffering of Microclimatic Extremes

Tree diversity also promotes microclimatic buffering, a largely untested mechanism correlated with vegetation structure (e.g. overstory cover, structural complexity, etc.) (Moreira and Abdala-Roberts 2026). For example, multi-layered vegetation reduces temperature extremes and stabilizes humidity (Chen et al. 1999; De Frenne et al. 2019), improving habitat quality for thermally sensitive taxa such as bats and birds. Cooler microclimates are known to favour forest specialist birds, particularly under warming and urbanization (Jankowski et al. 2013; Betts et al. 2018), while bat roosting and foraging activity are highly sensitive to temperature and humidity, with many species avoiding hot and dry conditions (Loeb and O'keefe 2006, Jung et al. 2012). However, evidence linking these abiotic controls to tree diversity is scant, but a recent study suggests this could be the case. Work by Coyne et al. (2025), it was found that bird occurrences (habitat use, abundance) and resulting impacts on herbivores increased with tree diversity and suggested that microclimatic variation driven by tree diversity shapes habitat use and foraging by insectivorous birds, either directly or indirectly via effects on insect prey availability (i.e. abiotic buffering effects of tree diversity on arthropods; see Schuldt et al. 2019). In the case of bats, this mechanism may be less relevant, because they are nocturnal and as a result usually forage under cooler conditions (albeit humidity levels can influence bat foraging and may vary with tree diversity). Still, tree diversity could play a role by mitigating adverse nocturnal microclimatic conditions, such as reducing air currents and buffering against low temperatures (e.g. in temperate forests). Even in tropical systems, diverse assemblages may influence nocturnal humidity levels, potentially affecting bat activity. Thus, the importance of microclimatic buffering likely varies across taxa and environments.

9.4.3 Changes in Prey and Tree-Based Resource Heterogeneity

More diverse forests are likely to provide a greater amount, diversity, and temporally more stable supply of prey for predators (Elton 1958; Root 1973; Lawton and Strong 1981). In addition, tree mixtures provide a greater heterogeneity of resources beyond insect prey, such as a wider variety of foraging substrates (Holmes and Schultz 1988) and complementary resources such as nectar, pollen, and fruits that enhance natural enemy fitness (Russell 1989; Cappuccino et al. 1999). Similarly, tree diversity increases the availability of nesting substrates and shelter, further promoting predator recruitment and residence time (Batáry et al. 2014; Asbeck et al. 2019). Comparable mechanisms could operate for bats, but formal tests are also lacking. Resource heterogeneity may affect bird and bat species differently depending on their degree of habitat and prey specialization. Generalist species may readily exploit diverse prey and roosting opportunities in mixed forests, whereas specialists with narrow habitat or dietary requirements may respond less strongly (May-Uc et al. 2020). Again, as for structural complexity and microclimatic variation, outcomes will be contingent on species traits and guild composition (see next section).

9.5 Context-Dependency in Tree Diversity Effects

We next explore variability in the strength, consistency, and direction of tree diversity-mediated feedbacks involving insectivorous birds and bats and discuss potential drivers of such context dependence.

9.5.1 Tree Identity and Community Composition

Tree species identity or composition may outweigh tree diversity effects on higher trophic levels. Schuldt et al. (2018) reviewed evidence showing that tree species identity and composition are the main drivers of insectivorous birds and bat communities at local scales in temperate forests (Penone et al. 2019; Dekeukeleire et al. 2019). Evidence from broader-scale studies also shows that bird communities often track the identity of specific tree species or assemblages rather than richness per se (Lee and Rotenberry 2005; Jankowski et al. 2013). Within this context, a recurrent biogeographic trend in temperate forest and Mediterranean systems is for deciduous broadleaved forests (e.g., oak-dominated stands) to host richer and more specialized bird communities than coniferous forests. For example, bird richness tends to decline with increasing conifer cover (Gil-Tena et al. 2007), and mixed oak–pine stands support more species than pure pine forests, with oaks acting as key drivers (Díaz 2006). A plausible explanation for the role of tree species composition relies on chemical or structural traits (reviewed above) of focal tree species or specific assemblages that shape associated herbivore and predator assemblages (Singer et al. 2012; Abdala-Roberts et al. 2019). In one of the few available tree diversity studies hinting at this, Castagneyrol et al. (2017) demonstrated that functionally dissimilar tree mixtures amplified bird attack rates on sentinel prey and reduced herbivory, whereas functionally redundant mixtures did not, underscoring tree composition controls over top-down forcing.

9.5.2 Insect Prey Traits and Population Responses

Tree diversity effects on insectivorous vertebrates are expected to be strongly dependent on prey community composition. In particular, herbivore functional traits, which reflect species composition, are likely to differentially influence different insectivorous vertebrate species or guilds responses to tree diversity and, in turn, their aggregate top-down effects. For instance, tree diversity is predicted to negatively affect the abundance of specialist herbivores and positively affect generalist herbivores (Jactel et al. 2021), with these shifts in insect relative abundances of each group in turn affecting predator recruitment and predation pressure in mixed tree stands if prey preference is structured by herbivore diet breadth. Namely, lower recruitment and foraging activity in tree mixtures by insectivores that feed on specialist herbivores (further strengthened by crypsis or aposematism in these taxa; Singer et al. 2014), while the reverse might hold true for those feeding on generalist herbivores. Other prey traits such as feeding mode may also influence trophic cascades. Internal-feeding herbivores, such as leaf miners or borers, can occupy enemy-free space, reducing their detectability or accessibility to vertebrate predators, whereas external feeders, such as leaf chewers or sap feeders, are generally more vulnerable (Staab and Schuldt 2020). Here, shifts toward internal-feeding herbivores in tree mixtures may dampen tree diversity-associated trophic cascades.

It is important to also consider insect prey population responses to host plant availability. Combined with diet breadth, for example, generalist herbivores responding positively to plant diversity (Salazar et al. 2013; Zhang et al. 2017) may exhibit positive density-dependent recruitment to mixtures and reach outbreaks levels that overwhelm insectivorous vertebrates (Nixon and Roland 2012). Even if highly efficient at detecting host plants, this outcome would be less likely for specialist herbivores given expected lower recruitment and overall abundances in tree mixtures by these insect species or guilds. Lastly, herbivore dispersal capacity may interact with tree diversity, synergistically contributing to positive density-dependent recruitment and predator satiation in mixtures (Hambäck et al. 2014).

9.5.3 Predator Identity and Predator-Predator Interactions

Insectivorous vertebrate communities are functionally heterogeneous, exhibiting species and guild differences in prey preferences (see previous section), as well as in other important traits such mobility, sensory modalities, prey-capture strategies, and abiotic tolerance (see Sect. 9.2), all of which may influence vertebrate responses to tree diversity and their top-down effects. For instance, Vázquez-González et al. (2024b) found that omnivorous birds responded positively and more strongly to mixtures than specialized insectivores, suggesting that dietary breadth filter responses to tree diversity and potential top-down effects. Generalist omnivore consumers might benefit more from diverse food resources in mixed stands, while specialists (e.g. obligate insectivores) may be less responsive (Zhang et al. 2017; Staab and Schuldt 2020). Relatedly, predator migratory status may also be of predictive use. For instance, migratory birds often show narrower, insectivorous diets (Greenberg 1995; Van Bael et al. 2008; May-Uc et al. 2020) and broader abiotic tolerances (Cohen et al. 2020). As a result, migratory birds may benefit less from increased prey diversity or microclimatic buffering in diverse forests compared to resident species. A similar framework can be extended to compare birds and bats, as both groups differ in sensory systems, mobility, and foraging specialization.

Tree diversity effects on trophic cascades are further shaped by interactions among predator guilds (reviewed by Abdala-Roberts and Moreira 2026). These interactions can either amplify or dampen diversity effects depending on whether tree diversity promotes niche partitioning among predators (e.g. via specialization on different prey or physical refugia) or antagonism through increased predator interference or competition as a result of higher predator diversity (Staab and Schuldt 2020). On the one hand, structurally complex forests can promote coexistence of functionally complementary guilds (e.g. volant predators such as birds and bats often partition vertical strata and diel activity, boosting top-down control of herbivores). However, negative interactions may counteract such gains. Although evidence in forests is very limited, there is some indication that this could be the case. For example, Interian-Aguiñaga et al. (2022) found negative correlations between bird and predatory arthropod attack rates on dummy caterpillars in tree mixtures, suggesting tree diversity strengthened negative interactions between these groups. We are not aware of studies showing negative (e.g. competitive) interactions between birds and bats, though behavioural and foraging differences (e.g. diurnal vs. nocturnal foraging and associated prey items) may instead foster complementarity. Lastly, interactions between carnivorous vertebrates and insectivorous vertebrates (e.g. insectivorous birds and raptors) should also be considered but are virtually unstudied.

9.5.4 Landscape-Level and Broader-Scale Dynamics

Tree diversity effects on predators and their cascading impacts are strongly shaped by spatial variation in abiotic and biotic factors, which modulate tree diversity–function relationships across scales. For example, macroclimatic variation can influence the strength of tree diversity effects. Positive effects are expected to be stronger under harsher conditions, such as elevated temperatures or frequent droughts, because diverse forests provide microclimatic refugia for sensitive species and buffer abiotic stress more effectively in these environments (Betts et al. 2018; De Frenne et al. 2019). Supporting this idea, an observational study reported that tree-species richness effects on vertebrate communities were at least partially explained by climate (Ampoorter et al. 2020). In contrast, Vázquez-González et al. (2024a) reported that diversity–predation relationships were consistent across broad climatic gradients in *TreeDivNet* experiments, suggesting that large-scale climate does not always modulate tree diversity effects. At mesoscales, on the other hand, landscape context also varies at large-scale and further shapes outcomes. Local diversity effects are embedded within broader patterns of forest cover, connectivity, and fragmentation (Martin et al. 2013). For example, in agroforestry systems in North America, Mitchell et al. (2014) showed that high forest cover and connectivity strengthened top-down control, whereas fragmentation weakened the ability of tree diversity to enhance predation.

Seasonal cycles, interannual fluctuations, and long-term climatic variation in both biotic and abiotic factors can also shape tree diversity effects on insectivorous vertebrates. For birds (and potentially also bats), the influence of migration patterns is important but remains poorly studied. At the UADY-UNAM *TreeDivNet* system in Mexico, resident birds responded more strongly to tree diversity than migratory species (May-Uc et al. 2020; Vázquez-González et al. 2024b), possibly because migrants, constrained by fixed arrival and breeding schedules, are less able to track resource pulses in diverse stands and are less sensitive to changes in prey diversity or microclimatic conditions. In another study, however, it was found that migratory insectivores can exert disproportionately strong top-down control on herbivores (Van Bael et al. 2008). Regardless, the timing and degree of migrant influx could shape community-wide effects and contribute to seasonal variation in top-down control by insectivorous vertebrates. Similarly, temporal changes in prey communities can contribute to variation in tree diversity effects. Here, seasonal fluctuations in prey life cycles and guild composition alter the match between predators and resources (Southwood et al. 2004; Pureswaran et al. 2018). For example, bird breeding seasons often coincide with caterpillar population peaks, but mismatches can reduce suppression potential (Burgess et al. 2018). Accordingly, a focus on these dynamics from the perspective of abiotic variability driving prey or directly influencing predators, including seasonal and interannual changes in temperature and humidity, as well as longer-scale climatic variation, is key to consider in testing for tree diversity effects on insectivorous vertebrates. For example, tree diversity effects may be amplified during harsh periods (e.g. dry seasons or drought years) through stronger microclimatic buffering, as discussed for spatial variation. Yet evidence remains inconclusive: While May-Uc et al. (2020) discussed seasonal differences in tree diversity effects on bird diversity as a result of shifts in climatic conditions, resource availability, and migratory species influxes, a study by Vázquez-González et al. (2024b) in the same system (UADY-UNAM Tree Diversity Experiment) found no clear trend for seasonal variation in bird predatory activity, suggesting links between tree diversity-driven changes in bird community structure and predation rates are not always straightforward. Overall, these findings highlight that temporal variation in tree diversity effects is likely shaped by the interplay of biotic and abiotic interrelated conditions, yet to be teased a part via mechanistic studies involving longer-term measurements.

9.6 Concluding Remarks

Tree diversity emerges as a central driver of insectivorous bird and bat communities, with this having key conservation and management implications. Although still at its infancy, research from local-scale manipulations of tree diversity has provided some enticing indications for mechanisms such as microclimatic buffering and structural complexity behind tree diversity effects on these predator groups. Yet, tree diversity effects have been shown to vary in strength (and even direction), and the sources behind this variability are yet to be formally tested and revolve around a better understanding of tree, prey, and predator traits, spatiotemporal variation in predator-prey interactions, abiotic factors, and the relationships between local-scale diversity effects and broader-scale variation in biotic and abiotic factors. Future research should prioritize experimental approaches involving selective predator exclusions, more work to better understand insectivorous bat responses, longer-term measurements, as well as studies focusing on other groups such as rodents and reptiles. In doing so, we can better inform forest management strategies that leverage knowledge on the linkages between tree diversity and multi-trophic interactions which sustain forest function and resilience.

References

Abdala-Roberts L, Moreira X (2026) Effects of plant diversity on predator-predator interactions in terrestrial ecosystems. In: Wajnberg E, Gontijo L (eds) The ecology of plant diversity. Springer

Abdala-Roberts L, Puentes A, Finke DL, Marquis RJ, Montserrat M, Poelman EH, Rasmann S, Sentis A, van Dam NM, Wimp G, Mooney K, Björkman C (2019) Tri-trophic interactions: bridging species, communities and ecosystems. Ecol Lett 22:2151–2167

Albert G, Gauzens B, Loreau M, Wang S, Brose U (2022) The hidden role of multi-trophic interactions in driving diversity–productivity relationships. Ecol Lett 25:405–415

Ampoorter E, Barbaro L, Jactel H, Baeten L, Boberg J, Carnol M, Castagneyrol B, Charbonnier Y, Dawud SM, Deconchat M, Smedt PD, Wandeler HD, Guyot V, Hättenschwiler S, Joly F-X, Koricheva J, Milligan H, Muys B, Nguyen D, Ratcliffe S, Raulund-Rasmussen K, Scherer-Lorenzen M, van der Plas F, Keer JV, Verheyen K, Vesterdal L, Allan E (2020) Tree diversity is key for promoting the diversity and abundance of forest-associated taxa in Europe. Oikos 129:133–146

Anttonen P, Perles-Garcia M, Kunz M, von Oheimb G, Li Y, Bruelheide H, Ma K-P, Zhu C-D, Schuldt A (2023) Predation pressure by arthropods, birds, and rodents is interactively shaped by tree species richness, vegetation structure, and season. Front Ecol Evol 11

Asbeck T, Pyttel P, Frey J, Bauhus J (2019) Predicting abundance and diversity of tree-related microhabitats in central European montane forests from common forest attributes. For Ecol Manag 432:400–408

Barber NA, Marquis RJ, Tori WP (2008) Invasive prey impacts the abundance and distribution of native predators. Ecology 89:2678–2683

Barton BT, Schmitz OJ (2009) Experimental warming transforms multiple predator effects in a grassland food web. Ecol Lett 12:1317–1325

Batáry P, Fronczek S, Normann C, Scherber C, Tscharntke T (2014) How do edge effect and tree species diversity change bird diversity and avian nest survival in Germany's largest deciduous forest? For Ecol Manag 319:44–50

Beilke EA, O'Keefe JM (2023) Bats reduce insect density and defoliation in temperate forests: an exclusion experiment. Ecology 104:e3903

Bereczki K, Ódor P, Csóka G, Mag Z, Báldi A (2014) Effects of forest heterogeneity on the efficiency of caterpillar control service provided by birds in temperate oak forests. For Ecol Manag 327:96–105

Betts MG, Phalan B, Frey SJ, Rousseau JS, Yang Z (2018) Old-growth forests buffer climate-sensitive bird populations from warming. Divers Distrib 24:439–447

Böhm SM, Wells K, Kalko EKV (2011) Top-down control of herbivory by birds and bats in the canopy of temperate broad-leaved oaks (Quercus robur). PLoS One 6:e17857

Bradbury RB, Hill RA, Mason DC, Hinsley SA, Wilson JD, Balzter H, Anderson GQ, Whittingham MJ, Davenport IJ, Bellamy PE (2005) Modelling relationships between birds and vegetation structure using airborne LiDAR data: a review with case studies from agricultural and woodland environments. Ibis 147:443–452

Brady MV, Farrer EC (2024) The soil microbiome affects patterns of local adaptation in an alpine plant under moisture stress. Am J Bot 111:e16304

Buchholz S, Kelm V, Ghanem SJ (2020) Mono-specific forest plantations are valuable bat habitats: implications for wind energy development. Eur J Wildl Res 67:1

Burgess MD, Smith KW, Evans KL, Leech D, Pearce-Higgins JW, Branston CJ, Briggs K, Clark JR, du Feu CR, Lewthwaite K, Nager RG, Sheldon BC, Smith JA, Whytock RC, Willis SG, Phillimore AB (2018) Tritrophic phenological match-mismatch in space and time. Nat Ecol Evol 2:970–975

Cappuccino N, Houle M-J, Stein J (1999) The influence of understory nectar resources on parasitism of the spruce budworm Choristoneura fumiferana in the field. Agric For Entomol 1:33–36

Castagneyrol B, Bonal D, Damien M, Jactel H, Meredieu C, Muiruri EW, Barbaro L (2017) Bottom-up and top-down effects of tree species diversity on leaf insect herbivory. Ecol Evol 7:3520–3531

Charbonnier Y, Gaüzère P, van Halder I, Nezan J, Barnagaud J-Y, Jactel H, Barbaro L (2016) Deciduous trees increase bat diversity at stand and landscape scales in mosaic pine plantations. Landsc Ecol 31:291–300

Chen J, Saunders SC, Crow TR, Naiman RJ, Brosofske KD, Mroz GD, Brookshire BL, Franklin JF (1999) Microclimate in forest ecosystem and landscape ecology: variations in local climate can be used to monitor and compare the effects of different management regimes. Bioscience 49:288–297

Cohen JM, Fink D, Zuckerberg B (2020) Avian responses to extreme weather across functional traits and temporal scales. Glob Chang Biol 26:4240–4250

Coyne S, Osorio E, Bennett SK, Kirchgraber A, Parker JD, Nowakowski AJ (2025) Bird occurrence and trophic interactions vary across gradients of tree diversity and microclimate in a planted forest. bioRxiv

De Frenne P, Zellweger F, Rodríguez-Sánchez F, Scheffers BR, Hylander K, Luoto M, Vellend M, Verheyen K, Lenoir J (2019) Global buffering of temperatures under forest canopies. Nat Ecol Evol 3:744–749

Dekeukeleire D, Hertzog LR, Vantieghem P, van Schrojenstein Lantman IM, Sercu BK, Boonyarittichaikij R, Martel A, Verheyen K, Bonte D, Strubbe D, Lens L (2019) Forest fragmentation and tree species composition jointly shape breeding performance of two avian insectivores. For Ecol Manag 443:95–105

Díaz L (2006) Influences of forest type and forest structure on bird communities in oak and pine woodlands in Spain. For Ecol Manag 223:54–65

Dodd LE, Lacki MJ, Britzke ER, Buehler DA, Keyser PD, Larkin JL, Rodewald AD, Wigley TB, Wood PB, Rieske LK (2012) Forest structure affects trophic linkages: how silvicultural disturbance impacts bats and their insect prey. For Ecol Manag 267:262–270

Elton CS (1958) The ecology of invasions by animals and plants. Springer Nature

Fayt P, Machmer MM, Steeger C (2005) Regulation of spruce bark beetles by woodpeckers – a literature review. For Ecol Manag 206:1–14

Felix JA, Koricheva J (2006) Role of tree traits in mediating effects of forest diversity on herbivores. In: Moreira X, Abdala-Roberts L, Vázquez-González C (eds) Effects of plant diversity on plant-herbivore interactions. Springer

Fujioka E, Aihara I, Sumiya M, Aihara K, Hiryu S (2016) Echolocating bats use future-target information for optimal foraging. Proc Natl Acad Sci 113:4848–4852

Gamfeldt L, Snäll T, Bagchi R, Jonsson M, Gustafsson L, Kjellander P, Ruiz-Jaen M, Fröberg M, Stendahl J, Philipson C (2013) Higher levels of multiple ecosystem services are found in forests with more tree species. Nat Commun 4

Gil-Tena A, Saura S, Brotons L (2007) Effects of forest composition and structure on bird species richness in a Mediterranean context: implications for forest ecosystem management. For Ecol Manag 242:470–476

Greenberg R (1995) Insectivorous migratory birds in tropical ecosystems: the breeding currency hypothesis. J Avian Biol 26:260–264

Grossman JJ, Cavender-Bares J, Reich PB, Montgomery RA, Hobbie SE (2019) Neighborhood diversity simultaneously increased and decreased susceptibility to contrasting herbivores in an early stage forest diversity experiment. J Ecol 107:1492–1505

Haddad NM, Crutsinger GM, Gross K, Haarstad J, Knops JM, Tilman D (2009) Plant species loss decreases arthropod diversity and shifts trophic structure. Ecol Lett 12:1029–1039

Hambäck PA, Inouye BD, Andersson P, Underwood N (2014) Effects of plant neighborhoods on plant–herbivore interactions: resource dilution and associational effects. Ecology 95:1370–1383

Hobson KA, Bayne E (2000) Breeding bird communities in boreal Forest of Western Canada: consequences of "Unmixing" the Mixedwoods. Condor 102:759–769

Holmes RT, Schultz JC (1988) Food availability for forest birds: effects of prey distribution and abundance on bird foraging. Can J Zool 66:720–728

Huang Y, Chen Y, Castro-Izaguirre N, Baruffol M, Brezzi M, Lang A, Li Y, Härdtle W, Von Oheimb G, Yang X (2018) Impacts of species richness on productivity in a large-scale subtropical forest experiment. Science 362:80–83

Interian-Aguiñaga J, Parra-Tabla V, Abdala-Roberts L (2022) Effects of topical tree diversity and prey spatial distribution on predation by birds and arthropods. Arthropod Plant Interact 16:449–457

Jactel H, Moreira X, Castagneyrol B (2021) Tree diversity and forest resistance to insect pests: patterns, mechanisms, and prospects. Annu Rev Entomol 66:277–296

Jankowski JE, Merkord CL, Rios WF, Cabrera KG, Revilla NS, Silman MR (2013) The relationship of tropical bird communities to tree species composition and vegetation structure along an Andean elevational gradient. J Biogeogr 40:950–962

Juchheim J, Ehbrecht M, Schall P, Ammer C, Seidel D (2020) Effect of tree species mixing on stand structural complexity. Forestry Int J Forest Res 93:75–83

Jung K, Kaiser S, Böhm S, Nieschulze J, Kalko EKV (2012) Moving in three dimensions: effects of structural complexity on occurrence and activity of insectivorous bats in managed forest stands. J Appl Ecol 49:523–531

Kalka MB, Smith AR, Kalko EKV (2008) Bats limit arthropods and herbivory in a tropical forest. Science 320:71–71

Karr JR (1976) Seasonality, resource availability, and community diversity in tropical bird communities. Am Nat 110:973–994

Kim H, McComb BC, Frey SJK, Bell DM, Betts MG (2022) Forest microclimate and composition mediate long-term trends of breeding bird populations. Glob Chang Biol 28:6180–6193

Kissling WD, Rahbek C, Böhning-Gaese K (2007) Food plant diversity as broad-scale determinant of avian frugivore richness. Proc R Soc B Biol Sci 274:799–808

Koenig WD, Liebhold AM, Bonter DN, Hochachka WM, Dickinson JL (2013) Effects of the emerald ash borer invasion on four species of birds. Biol Invasions 15:2095–2103

Kollberg I, Bylund H, Huitu O, Björkman C (2014) Regulation of forest defoliating insects through small mammal predation: reconsidering the mechanisms. Oecologia 176:975–983

Kremer K, Jonsson B-G, Dutta T, Tavares MF, Bauhus J (2025) Single- vs mixed-species plantations: a systematic review on the effects on biodiversity. Biol Conserv 307:111182

Kunz TH, Braun de Torrez E, Bauer D, Lobova T, Fleming TH (2011) Ecosystem services provided by bats. Ann N Y Acad Sci 1223:1–38

Langellotto GA, Denno RF (2004) Responses of invertebrate natural enemies to complex-structured habitats: a meta-analytical synthesis. Oecologia 139:1–10
Lawton JH, Strong DR (1981) Community patterns and competition in folivorous insects. Am Nat 118:317–338
Lee P-Y, Rotenberry JT (2005) Relationships between bird species and tree species assemblages in forested habitats of eastern North America. J Biogeogr 32:1139–1150
Letourneau DK, Armbrecht I, Rivera BS, Lerma JM, Carmona EJ, Daza MC, Escobar S, Galindo V, Gutiérrez C, López SD, Mejía JL, Rangel AMA, Rangel JH, Rivera L, Saavedra CA, Torres AM, Trujillo AR (2011) Does plant diversity benefit agroecosystems? A synthetic review. Ecol Appl 21:9–21
Li Y, Schmid B, Schuldt A, Li S, Wang M-Q, Fornoff F, Staab M, Guo P-F, Anttonen P, Chesters D, Bruelheide H, Zhu C-D, Ma K, Liu X (2023) Multitrophic arthropod diversity mediates tree diversity effects on primary productivity. Nat Ecol Evol 7:832–840
Loeb SC, O'keefe JM (2006) Habitat use by forest bats in South Carolina in relation to local, stand, and landscape characteristics. J Wildl Manag 70:1210–1218
Low PA, Sam K, McArthur C, Posa MRC, Hochuli DF (2014) Determining predator identity from attack marks left in model caterpillars: guidelines for best practice. Entomol Exp Appl 152:120–126
Maas B, Karp DS, Bumrungsri S, Darras K, Gonthier D, Huang JC-C, Lindell CA, Maine JJ, Mestre L, Michel NL, Morrison EB, Perfecto I, Philpott SM, Şekercioğlu ÇH, Silva RM, Taylor PJ, Tscharntke T, Van Bael SA, Whelan CJ, Williams-Guillén K (2016) Bird and bat predation services in tropical forests and agroforestry landscapes. Biol Rev 91:1081–1101
MacArthur RH, MacArthur JW (1961) On bird species diversity. Ecology 42:594–598
Martin EA, Reineking B, Seo B, Steffan-Dewenter I (2013) Natural enemy interactions constrain pest control in complex agricultural landscapes. Proc Natl Acad Sci 110:5534–5539
Matlock RB Jr, Edwards PJ (2006) The influence of habitat variables on bird communities in forest remnants in Costa Rica. Biodivers Conserv 15:2987–3016
May-Uc Y, Nell CS, Parra-Tabla V, Navarro J, Abdala-Roberts L (2020) Tree diversity effects through a temporal lens: implications for the abundance, diversity and stability of foraging birds. J Anim Ecol 89:1775–1787
McCracken GF, Westbrook JK, Brown VA, Eldridge M, Federico P, Kunz TH (2012) Bats track and exploit changes in insect pest populations
Mitchell MGE, Bennett EM, Gonzalez A (2014) Agricultural landscape structure affects arthropod diversity and arthropod-derived ecosystem services. Agric Ecosyst Environ 192:144–151
Mooney KA, Linhart YB (2006) Contrasting cascades: insectivorous birds increase pine but not parasitic mistletoe growth. J Anim Ecol 75:350–357
Mooney KA, Gruner DS, Barber NA, Van Bael SA, Philpott SM, Greenberg R (2010) Interactions among predators and the cascading effects of vertebrate insectivores on arthropod communities and plants. Proc Natl Acad Sci 107:7335–7340
Moreira X, Abdala-Roberts L (2026) Effects of plant diversity on ants: a key but overlooked driver in multi-trophic interactions. In: Moreira X, Abdala-Roberts L, Vázquez-González C (eds) Effects of plant diversity on plant-herbivore interactions. Springer
Moreira X, Abdala-Roberts L, Rasmann S, Castagneyrol B, Mooney KA (2016) Plant diversity effects on insect herbivores and their natural enemies: current thinking, recent findings, and future directions. Curr Opin Insect Sci 14:1–7
Muiruri EW, Rainio K, Koricheva J (2016) Do birds see the forest for the trees? Scale-dependent effects of tree diversity on avian predation of artificial larvae. Oecologia 180:619–630
Nell CS, Abdala-Roberts L, Parra-Tabla V, Mooney KA (2018) Tropical tree diversity mediates foraging and predatory effects of insectivorous birds. Proc R Soc B 285:20181842
Nixon AE, Roland J (2012) Generalist predation on forest tent caterpillar varies with forest stand composition: an experimental study across multiple life stages. Ecol Entomol 37:13–23
Paquette A, Hector A, Castagneyrol B, Vanhellemont M, Koricheva J, Scherer-Lorenzen M, Verheyen K (2018) A million and more trees for science. Nat Ecol Evol 2:763–766

Penone C, Allan E, Soliveres S, Felipe-Lucia MR, Gossner MM, Seibold S, Simons NK, Schall P, van der Plas F, Manning P, Manzanedo RD, Boch S, Prati D, Ammer C, Bauhus J, Buscot F, Ehbrecht M, Goldmann K, Jung K, Müller J, Müller JC, Pena R, Polle A, Renner SC, Ruess L, Schönig I, Schrumpf M, Solly EF, Tschapka M, Weisser WW, Wubet T, Fischer M (2019) Specialisation and diversity of multiple trophic groups are promoted by different forest features. Ecol Lett 22:170–180

Perfecto I, Vandermeer J, Philpott SM (2014) Complex ecological interactions in the coffee agroecosystem. Annu Rev Ecol Evol Syst 45:137–158

Pianka ER, Vitt LJ, Pelegrin N, Fitzgerald DB, Winemiller KO (2017) Toward a periodic table of niches, or exploring the lizard niche hypervolume. Am Nat 190:601–616

Poeydebat C, Jactel H, Moreira X, Koricheva J, Barsoum N, Bauhus J, Eisenhauer N, Ferlian O, Francisco M, Gottschall F, Gravel D, Mason B, Muiruri E, Muys B, Nock C, Paquette A, Ponette Q, Scherer-Lorenzen M, Stokes V, Staab M, Verheyen K, Castagneyrol B (2021) Climate affects neighbour-induced changes in leaf chemical defences and tree diversity–herbivory relationships. Funct Ecol 35:67–81

Preisser EL, Bolnick DI, Benard MF (2005) Scared to death? The effects of intimidation and consumption in predator–prey interactions. Ecology 86:501–509

Pureswaran DS, Roques A, Battisti A (2018) Forest insects and climate change. Curr For Rep 4:35–50

Ratcliffe S, Wirth C, Jucker T, van Der Plas F, Scherer-Lorenzen M, Verheyen K, Allan E, Benavides R, Bruelheide H, Ohse B (2017) Biodiversity and ecosystem functioning relations in European forests depend on environmental context. Ecol Lett 20:1414–1426

Richter H, Cumming G (2008) First application of satellite telemetry to track African straw-coloured fruit bat migration. J Zool 275:172–176

Root RB (1973) Organization of a plant-arthropod association in simple and diverse habitats: the fauna of collards (Brassica oleracea). Ecol Monogr 43:95–124

Rubene D, Leidefors M, Ninkovic V, Eggers S, Low M (2019) Disentangling olfactory and visual information used by field foraging birds. Ecol Evol 9:545–552

Russell EP (1989) Enemies hypothesis: a review of the effect of Vegetational diversity on predatory insects and parasitoids. Environ Entomol 18:590–599

Salazar D, Kelm DH, Marquis RJ (2013) Directed seed dispersal of piper by Carollia perspicillata and its effect on understory plant diversity and folivory. Ecology 94:2444–2453

Schillé L, Paquette A, Marcotte G, Ouellet H, Cobus S, Barbaro L, Castagneyrol B (2025) Urban tree diversity fosters bird insectivory despite a loss in bird diversity with urbanization. Landsc Urban Plan 256:105274

Schmitz OJ, Beckerman AP, O'Brien KM (1997) Behaviorally mediated trophic cascades: effects of predation risk on food web interactions. Ecology 78:1388–1399

Schnitzler H-U, Kalko EK (2001) Echolocation by insect-eating bats: we define four distinct functional groups of bats and find differences in signal structure that correlate with the typical echolocation tasks faced by each group. Bioscience 51:557–569

Schnitzler H-U, Moss CF, Denzinger A (2003) From spatial orientation to food acquisition in echolocating bats. Trends Ecol Evol 18:386–394

Schuldt A, Assmann T, Brezzi M, Buscot F, Eichenberg D, Gutknecht J, Härdtle W, He J, Klein A, Kühn P (2018) Biodiversity across trophic levels drives multifunctionality in highly diverse forests. Nat Commun 9:2989

Schuldt A, Ebeling A, Kunz M, Staab M, Guimarães-Steinicke C, Bachmann D, Buchmann N, Durka W, Fichtner A, Fornoff F (2019) Multiple plant diversity components drive consumer communities across ecosystems. Nat Commun 10:1460

Sekercioglu CH (2012) Bird functional diversity and ecosystem services in tropical forests, agroforests and agricultural areas. J Ornithol 153:153–161

Sherry TW (1984) Comparative dietary ecology of sympatric, insectivorous Neotropical flycatchers (Tyrannidae). Ecol Monogr 54:313–338

Singer MS, Farkas TE, Skorik CM, Mooney KA (2012) Tritrophic interactions at a community level: effects of host plant species quality on bird predation of caterpillars. Am Nat 179:363–374

Singer MS, Lichter-Marck IH, Farkas TE, Aaron E, Whitney KD, Mooney KA (2014) Herbivore diet breadth mediates the cascading effects of carnivores in food webs. Proc Natl Acad Sci 111:9521–9526

Sivault E, Koane B, Chmurova L, Sam K (2024) Birds and bats reduce herbivory damage in Papua New Guinean highland forests. Ecology 105:e4421

Southwood TRE, Wint GW, Kennedy CE, Greenwood SR (2004) Seasonality abundance, species richness and specificity of the phytophagous guild of insects on oak (Quercus) canopies. Eur J Entomol 101:43–50

Staab M, Schuldt A (2020) The influence of tree diversity on natural enemies – a review of the "enemies" hypothesis in forests. Curr For Rep 6:243–259

Stemmelen A, Jactel H, Brockerhoff E, Castagneyrol B (2022a) Meta-analysis of tree diversity effects on the abundance, diversity and activity of herbivores' enemies. Basic Appl Ecol 58:130–138

Stemmelen A, Paquette A, Benot M-L, Kadiri Y, Jactel H, Castagneyrol B (2022b) Insect herbivory on urban trees: complementary effects of tree neighbours and predation. Peer Commun J 2

Tilman D, Isbell F, Cowles JM (2014) Biodiversity and ecosystem functioning. Annu Rev Ecol Evol Syst 45:471–493

Van Bael SA, Brawn JD, Robinson SK (2003) Birds defend trees from herbivores in a neotropical forest canopy. Proc Natl Acad Sci 100:8304–8307

Van Bael SA, Philpott SM, Greenberg R, Bichier P, Barber NA, Mooney KA, Gruner DS (2008) Birds as predators in tropical agroforestry systems. Ecology 89:928–934

van Der Plas F, Manning P, Soliveres S, Allan E, Scherer-Lorenzen M, Verheyen K, Wirth C, Zavala MA, Ampoorter E, Baeten L (2016) Biotic homogenization can decrease landscape-scale forest multifunctionality. Proc Natl Acad Sci 113:3557–3562

Vázquez-González C, Castagneyrol B, Muiruri EW, Barbaro L, Abdala-Roberts L, Barsoum N, Fründ J, Glynn C, Jactel H, McShea WJ, Mereu S, Mooney KA, Morillas L, Nock CA, Paquette A, Parker JD, Parker WC, Roales J, Scherer-Lorenzen M, Schuldt A, Verheyen K, Weih M, Yang B, Koricheva J (2024a) Tree diversity enhances predation by birds but not by arthropods across climate gradients. Ecol Lett 27:e14427

Vázquez-González C, Villa-Galaviz E, Reyes-Hernández M, Perez-Niño B, Quijano-Medina T, Parra-Tabla V, Mooney KA, Abdala-Roberts L (2024b) Temporal variation in tree diversity effects on birds and its implications for top-down control of insect herbivores in a tropical system. Oecologia 204:603–612

Whelan CJ (2001) Foliage structure influences foraging of insectivorous forest birds: an experimental study. Ecology 82:219–231

Whelan CJ, Wenny DG, Marquis RJ (2008) Ecosystem services provided by birds. Ann N Y Acad Sci 1134:25–60

Yang B, Li B, He Y, Zhang L, Bruelheide H, Schuldt A (2018) Tree diversity has contrasting effects on predation rates by birds and arthropods on three broadleaved, subtropical tree species. Ecol Res 33:205–212

Zhang J, Bruelheide H, Chen X, Eichenberg D, Kröber W, Xu X, Xu L, Schuldt A (2017) Tree diversity promotes generalist herbivore community patterns in a young subtropical forest experiment. Oecologia 183:455–467

Zvereva EL, Zverev V, Kozlov MV (2020) Predation and parasitism on herbivorous insects change in opposite directions in a latitudinal gradient crossing a boreal forest zone. J Anim Ecol 89:2946–2957

Synthesis: Plant Diversity, Herbivory, and Ecosystem Function

This volume brings together a broad and integrative body of research aimed at advancing our understanding of how plant biodiversity shapes herbivory and, ultimately, ecosystem functioning. Across diverse ecosystems and methodological approaches, the chapters collectively reinforce the importance of plant diversity as a central driver of multitrophic interactions. At the same time, they reveal that the effects of plant diversity on herbivory are far from uniform and emerge from complex interactions among plant traits, herbivores, predators, and environmental conditions. By synthesizing these perspectives, this volume contributes to a more nuanced and mechanistic understanding of biodiversity–ecosystem functioning relationships.

A consistent pattern emerging from multiple chapters is the importance of bottom-up processes in structuring herbivore communities. Increased plant diversity often translates into greater heterogeneity in resource availability, both in terms of quantity and quality. This can influence herbivore abundance, diversity, and feeding behaviour, depending on both plant traits and herbivore traits such as dietary specialization, dispersal ability, and feeding mode. In particular, plant trait-based approaches, especially those examining how plant diversity shapes host plant traits, can help illuminate how different plant–herbivore interactions respond to plant diversity and yield a predictive understanding of interaction outcomes at both focal plant and community levels. Accordingly, one of the most important insights highlighted in this synthesis is the variability in how plant diversity affects plant traits and, consequently, herbivory (Chap. 2). Surprisingly, the magnitude and direction of diversity effects vary widely across different types of traits. In some cases, increased plant diversity enhances defensive traits, such as the production of secondary metabolites or structural defences, leading to reduced herbivore damage. In other cases, however, diversity has neutral or even positive effects on herbivory. This inconsistency reflects that plant responses to diversity are often species-specific and trait-dependent. Moreover, changes in plant traits do not always translate directly into herbivore responses (abundance or herbivory rates), suggesting that additional mechanisms are at play.

X. Moreira et al. (eds.), *Effects of Plant Diversity on Plant-Herbivore Interactions*, Ecological Studies 255, https://doi.org/10.1007/978-3-032-26461-9

Several explanations may account for this mismatch. First, methodology-related factors can play a role. The timing of trait measurements relative to herbivore activity is critical, as plant traits can vary seasonally or in response to herbivory (see also Chap. 4 on other temporal perspectives). Sampling at phenologically inappropriate times may fail to capture relevant dynamics as plant–herbivore interactions unfold over the growing season. Second, the choice of traits measured may not always align with those most important for herbivores. For instance, focusing exclusively on chemical defences may overlook the role of physical traits such as leaf toughness, trichomes, or structural complexity. Additionally, an emergent view from this chapter is the possibility that different plant traits exert opposing effects on herbivores. For example, high nutritional content may attract herbivores, while high levels of defensive compounds deter them. These trade-offs may obscure the interpretation of patterns and are not apparent unless multiple plant traits are measured. Within this context, the role of growth–defence trade-offs in mediating plant diversity effects also remains an important but underexplored area. Understanding how plants allocate resources between growth and defence along gradients of plant species diversity or composition is likely to contribute to explaining how plant functional (trait) diversity is shaped by community context and, in turn, exerts controls over herbivory, which may further feedback to affect plant communities. Accordingly, trait-based approaches provide a promising avenue for addressing these research gaps. By focusing on functional traits rather than species identity, researchers can better link plant diversity to ecological processes. As emphasized in several chapters, future studies should move beyond observational approaches and explicitly manipulate trait diversity. Selecting plant species based on known functional traits or life-history strategies (e.g. acquisitive versus conservative strategies) could help clarify the mechanisms underlying diversity effects on herbivory. Furthermore, distinguishing between traits that are highly responsive to abiotic conditions (e.g. related to Chap. 6) and those that are more stable could help elucidate interactions between plant diversity and abiotic factors.

Several chapters highlight the importance of top-down control in shaping plant diversity–structured herbivore dynamics and plant productivity. The Enemies Hypothesis, which posits that increased plant diversity enhances the abundance and effectiveness of natural enemies, is partially or fully supported in several systems (e.g. Chaps. 3 and 5). Diverse plant communities can provide more resources, habitats, and microclimatic conditions for predators, leading to increased predation pressure on herbivores. Importantly, this volume extends the applicability of the Enemies Hypothesis beyond well-studied grassland systems to include temperate and tropical forests, as well as urban ecosystems (Chap. 7). In addition, and despite the generality of top-down processes, several chapters also highlight important context dependencies. For instance, the strength of predator-mediated control can vary depending on factors such as predator identity, community structure, environmental conditions, and other plant community structural features. Here, Chaps. 8 and 9 draw particular attention to the role of underrepresented predator groups, such as ants and insectivorous vertebrates, including birds and bats. These organisms can exert strong top-down effects and are often highly responsive to changes in plant diversity. Incorporating these groups into future research is essential for developing a more complete understanding of trophic interactions. Likewise, predator functional diversity within particular groups or guilds is likely to play a key role in determining the strength and stability of top-down control. Different predator species may vary in their hunting strategies, temporal activity patterns, and prey preferences. These differences can lead to synergy or antagonism among predators, influencing overall ecosystem functioning via trophic control. In this sense, studies that simultaneously examine multiple predator groups can provide valuable insights into these dynamics. There is also a need for more experimental approaches that directly test predator effects and their relative importance compared to bottom-up processes shaping herbivory and plant productivity. Predator-exclusion experiments across gradients of plant diversity, for example, can help disentangle causal relationships and quantify the contribution of predators to herbivore suppression and plant growth or quality. Moreover, long-term studies are particularly important for capturing temporal variability (see Chaps. 4, 5, and 6) and assessing the stability of these interactions over time.

Another key theme emerging from this volume is the importance of abiotic context, including disturbance regimes and global change drivers (e.g. see Chaps. 5 and 6). Plant diversity effects are often contingent on the abiotic environment, which can influence plant traits, herbivore performance, and predator activity. For instance, in systems shaped by recurrent disturbances such as fire or grazing, plant diversity is closely linked to these processes. Changes in disturbance regimes can alter community composition and disrupt trophic interactions. Global change drivers, such as climate warming, altered precipitation patterns, nutrient enrichment, and elevated atmospheric CO_2, are also proving to be important in predicting plant diversity effects. Evidence reviewed in several chapters is consistent with the premise that plant diversity can buffer ecosystems against these stressors, enabling them to maintain ecosystem functioning in the face of altered environmental conditions. For example, diverse plant communities may be better able to sustain productivity under drought conditions or to stabilize nutrient cycling under increased nutrient inputs (cite chapters). However, the mechanisms underlying these buffering effects are not fully understood. In particular, there is a need to examine how plant and consumer traits mediate the interactive effects of plant diversity and global change on herbivory (as highlighted in Chap. 6). Changes in plant chemistry under elevated CO_2, for instance, may affect herbivore feeding behaviour and predator–prey interactions, with cascading effects on ecosystem processes. Factorial experiments that manipulate both plant diversity and abiotic factors are essential for disentangling these complex interactions. Importantly, urban ecosystems represent an emerging frontier within these varying abiotic contexts (Chap. 7). Urban environments are characterized by novel species assemblages, altered disturbance regimes, and strong environmental gradients. These conditions provide unique opportunities to test ecological theory and examine how plant diversity influences herbivory and ecosystem functioning in human-dominated landscapes strongly shaped by abiotic factors acting at local to mesoscales. Understanding these dynamics is particularly important given the increasing extent of urbanization worldwide and the need to design green spaces that support biodiversity and ecosystem services.

Overall, the chapters in this volume highlight the importance of integrating multiple perspectives to better understand plant diversity effects and inform habitat management strategies in both natural and human-dominated (including urban) ecosystems. Concretely, trait-based approaches provide insights into the mechanisms linking plant diversity to ecological processes, while multitrophic analyses reveal how altered trophic controls shape herbivory and ecosystem function. At the same time, there is a need to consider the abiotic context shaping plant traits and multitrophic interactions.

Integrating Perspectives Across Scales

As a whole, the content of this volume ultimately points to integrating trait-based, multitrophic, and abiotic perspectives as the necessary step forward to advance biodiversity research. Trait-based approaches allow researchers to move beyond species counts and focus on the functional characteristics that influence ecological interactions. Chemical, nutritional, and morphological plant traits affect herbivores and predators within diverse plant communities, and these dynamics are further altered by environmental conditions acting on plant diversity–plant trait and plant trait–consumer linkages.

In doing so, long-term or longitudinal studies will be essential. Many ecological processes are inherently dynamic, varying across seasons and years. Phenological mismatches between plants and herbivores, temporal complementarity among species, and delayed responses to environmental change can all influence ecosystem functioning. Longitudinal studies are therefore essential for understanding how plant diversity effects unfold over time and predicting long-term effects acting via bottom-up and/or top-down processes (e.g. within- and among-season variation in plant diversity effects on plant traits or predator top-down effects in relation to herbivore abundance or outbreaks). Collaborative research networks and standardized methodologies offer valuable opportunities to address these challenges at larger spatial and temporal scales. In turn, by combining data across ecosystems and regions (e.g. TreeDivNet), researchers can identify general patterns while accounting for system-specific variation. Such efforts are particularly important for predicting ecosystem responses to global change and for informing management and conservation strategies.

From an applied perspective, the integration of these approaches across temporal and spatial scales has important implications for ecosystem management and restoration. Enhancing plant diversity—particularly functional diversity—can promote more resilient ecosystems by designing systems that harness regulatory processes, either bottom-up or top-down, underlying key ecosystem services such as pest control and plant biomass production (for timber or plant foods). Similarly, incorporating knowledge of plant traits and trophic interactions into restoration design can improve initial establishment and long-term success, combined with long-term monitoring of species interactions and ecosystem functions.

Future Directions

Despite substantial progress, several key gaps remain in understanding plant diversity and its ecological effects. First, mechanistic studies linking specific plant traits to herbivory are needed, as manipulating trait diversity directly can reveal causal relationships more clearly than species-level approaches. Second, multitrophic interactions remain underexplored; integrating multiple predator groups, including ants, birds, and bats, will clarify how predator diversity shapes ecosystem functioning. Third, abiotic factors such as temperature, moisture, and nutrients should be incorporated into experimental designs to disentangle their interactive effects with plant diversity, ideally across environmental gradients. Fourth, long-term studies are essential to capture temporal dynamics and assess ecosystem stability, particularly in forests where processes operate over decades. Fifth, urban ecosystems provide underutilized opportunities to examine biodiversity–function relationships in human-dominated landscapes. Finally, coordinated, large-scale research networks and standardized protocols are crucial for improving replication, comparability, and generalization across systems.